Student Solutions Manual

for

Introduction to Regression Modeling

Bovas Abraham
University of Waterloo

Johannes Ledolter
University of Iowa

BROOKS/COLE
CENGAGE Learning™

Australia • Brazil • Japan • Korea • Mexico • Singapore • Spain • United Kingdom • United States

BROOKS/COLE
CENGAGE Learning™

Student Solutions Manual for
Introduction to Regression Modeling
Bovas Abraham and Johannes Ledolter

For product information and
technology assistance, contact us at **Cengage Learning
Customer & Sales Support, 1-800-354-9706**

For permission to use material from this text or product,
submit all requests online at **www.cengage.com/permissions**
Further permissions questions can be e-mailed to
permissionrequest@cengage.com

ISBN-13: 978-0-534-42076-5
ISBN-10: 0-534-42076-1

Brooks/Cole Cengage Learning
20 Davis Drive
Belmont, CA 94002-3098
USA

Cengage Learning is a leading provider of customized learning solutions with office locations around the globe, including Singapore, the United Kingdom, Australia, Mexico, Brazil, and Japan. Locate your local office at **www.cengage.com/global**

Cengage Learning products are represented in Canada by Nelson Education, Ltd.

To learn more about Brooks/Cole, visit **www.cengage.com/brookscole**

Purchase any of our products at your local college store or at our preferred online store **www.cengagebrain.com**

Printed in the United States of America
2 3 4 5 6   23 22 21 20 19

Table of Contents

Acknowledgment: We would like to thank A. M. Variyath and J. E. Choi for helping with the solutions to several of the exercises in Chapters 2 and 4-7.

CHAPTER 1

1.1 Tensile strength of an alloy can be expected to increase with increasing hardness and density of the stock. Bivariate scatter plots of tensile strength against hardness and of tensile strength against density of the stock are useful. Such scatter plots indicate whether the relationship is linear, or more complicated.

Bivariate scatter plots are unable to reveal 3-dimensional relationships. For that one needs to consider three-dimensional graphs. Alternatively, one can proceed as follows. If measurements on the tensile strengths of several different alloys of a given density but of changing values of hardness are given, one can plot tensile strength against hardness at this one fixed level of density. Furthermore, if tensile strength and hardness data for alloys of a second different density are available, one can construct a similar scatter plot for that other level of density. If the two scatter plots (scatter plots of tensile strength against hardness, at the two different levels of density) show different slopes, then the effect of hardness on tensile strength depends on the level of density. The factors hardness and density of the stock are said to interact in their effect on tensile strength.

Data from experiments are usually more informative as one can control the conditions under which the experimental runs are carried out. Experimentation is probably not possible in case (f). The relative humidity conditions in the plant can not be varied according to a fixed experimental plan. Instead, one takes measurements in the plant on the relative humidity, and at the same time on the output (performance) of the process. A danger with such data is that the relative humidity in the plant may be affected by unknown factors that also affect the output. The root cause is not the humidity of the plant, but these other "lurking" variables.

1.2 The graph given below indicates a linear relationship between ln(Payout) and the product of interest rate and maturity, with an intercept that depends on the invested principal. Note that the linear model in the transformed variables fits perfectly.

This is expected from the model Payout = Pexp(RT). Taking the logarithm on both sides of the equation, leads to ln(Payout) = ln(P) + RT. The intercept changes with the logarithm of the invested principle; the regression coefficient of RT is one.

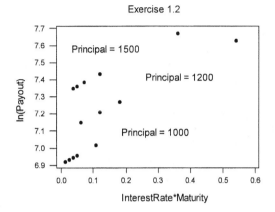

Exercise 1.2

1.3 Selected examples are:

Exercise 2.9: MBA grade point average and GMAT score: observational study
Exercise 2.10: Fuel efficiency and car characteristics: observational study of 45 cars
Exercise 2.24: Thickness of egg shell and PCB: observational study on pelicans
Exercise 2.27: Absorption of chemical liquid; experimental data
Exercise 4.12: Amount of plant water usage: observational study
Exercise 4.14: Survival of bull semen: experimental data
Exercise 4.15: Toxic action of a certain chemical on silkworm larvae: experimental data
Exercise 4.21: Abrasion as function of hardness and tensile strength of rubber:
            experimental data
Exercise 6.14: Tear properties of paper: experimental data
Exercise 6.17: Rigidity, elasticity and density of timber: observational study
Exercise 8.1: Incumbent vote share in US presidential elections: observational study
Exercise 8.2: Height and weight of boys: observational study
Exercise 8.3: Soft drink sales: observational study

1.6 Usually it is not very easy to spot relationships from 3-dimensional graphs; see the
two examples shown below. The bivariate scatter plots for the silkworm data set are
easier to interpret.

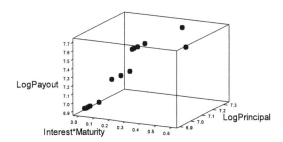

3-Dimensional Plot: Investment Data

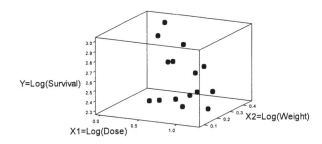

3-Dimensional Plot: Silkworm Data

1.7 Consider models with a single explanatory variable x. The quadratic model,
$$y = \beta_0 + \beta_1 x + \beta_2 x^2 + \varepsilon \ ,$$
is nonlinear in the explanatory variable x, but linear in the three parameters β_0, β_1 and β_2. The polynomial model (with $p > 1$),
$$y = \beta_0 + \beta_1 x + \beta_2 x^2 + ... + \beta_p x^p + \varepsilon,$$
is nonlinear in the explanatory variable x, but linear in the parameters.

The quadratic model with two explanatory variables,
$$y = \beta_0 + \beta_1 x_1 + \beta_2 x_2 + \beta_{11}(x_1)^2 + \beta_{22}(x_2)^2 + \beta_{12} x_1 x_2 + \varepsilon \ ,$$
is nonlinear in x_1 and x_2, however it is linear in the parameters. The equation describes a quadratic function in two variables. For certain values of the parameters the expected response looks like a bowl with a unique minimum, an upside bowl with a unique maximum, or a saddle point.

1.8 Consider a response y and a single explanatory variable x. The following models are nonlinear in the parameters. You may want to consider one of these models and trace out the mean response for changing levels of x. For example, take the first model with $\alpha = 0.39$ and $\beta = 0.10$ and consider x values between 8 and 40. This particular model is studied in Chapter 9; x is the age of a chemical product in weeks, and the response y is its remaining chlorine.

$$y = \alpha + (0.49 - \alpha)\exp[-\beta(x-8)] + \varepsilon$$

$$y = \beta_1 + \frac{\beta_2}{1 + \exp[-\beta_3(x - \beta_4]} + \varepsilon$$

$$y = \frac{\alpha}{1 + \beta\exp(-\gamma x)} + \varepsilon \quad \alpha > 0, \beta > 0, \gamma > 0$$

$$y = \frac{\beta_1 x}{\beta_2 + x} + \varepsilon$$

1.9 Sales may increase linearly with time, but the variability may depend on the level (the mean) of sales. If sales are very small, one can not expect tremendous variability. Sales can not be negative, so the variability is automatically bounded from below. On the other hand there is more room for bigger variability if the level of the sales is high. It is useful to think in terms of percentages. One may expect a variability (expressed as a standard deviation) of ± 10 percent. If sales are at level 10, this implies an uncertainty of ± 1 units. On the other hand, if the level is at 1000, the uncertainty is ± 100 units. If the variability (standard deviation) is proportional to the level, one should analyze the logarithm of sales, and not the sales. You will learn in Chapter 6 that this transformation stabilizes the variance. In this situation the variability in the logarithms of sales does not depend on the level of the sales.

Another situation, where the variability of the response can be expected to depend on the explanatory variable is when measuring distance. Assume that we want to determine the distances between pairs of points (where some are close together, while others are far apart). We can expect that the error in measuring close distances is smaller than the error in measuring points that are far apart. The variability in the measurements can be expected to increase with distance.

1.10 Economic "well-being" has an impact on people's decision to have children. During the post World War II period, a period characterized by rapid economic growth, many young Europeans affected by the war delayed their decision to have children. Economic activity of the post World War II period also had an impact on the breeding space for storks and led to a decrease in the number of storks. Considering annual numbers of births and annual numbers of storks, one can observe a strong positive correlation. However, no one - except young children - would interpret this correlation as a causal effect.

Poverty of a school district affects the number of students in subsidized lunch programs, with poorer districts having more children in these nationally subsidized programs. Poverty also affects the scholastic test scores in these districts. The strong positive correlation between the number of children in subsidized lunch programs and test achievement scores in these districts does not imply that there is a causal connection between subsidized lunch and test scores. It is poverty that is the driving causal factor.

High summer temperatures are related to high beer sales. High summer temperatures are also related to increased sales of suntan lotion. Daily sales of suntan lotion and beer sales are positively correlated. This, however, does not imply a causal connection. It is not that people who drink require more sun tan lotion.

1.12 (a) Ignoring variability, we find that for the ith subject: $\text{RelativeRaise}_i = \beta \text{Performance}_i$. All points in the graph of RelativeRaise against Performance are on a straight line through the origin.

The absolute raise (that is, the raise in terms of dollars earned) can be written as
$$\text{AbsoluteRaise} = (R)(\text{PreviousSalary}) = (\beta \text{PreviousSalary})\text{Performance}$$
A graph of AbsoluteRaise against Performance does not exhibit a perfect linear association as the slope depends on the previous salary that changes from person to person. A regression of AbsoluteRaise on Performance may not provide the correct estimate of the parameter β. Take two workers; the previous salary of the first worker is half the salary of the second one, but the first worker is twice as productive. Their absolute raises are the same. The slope in the plot of AbsoluteRaise against Performance is zero, and not the desired parameter β.

(b) Let R = Relative Raise, where R is a small number such as 0.03 (3 percent). The ratio $\text{CurrentSalary}/\text{PreviousSalary} = [(1 + R)\text{PreviousSalary}]/\text{PreviousSalary} = 1 + R$. A first-order Taylor series expansion of $\ln(1 + R) \approx R$ is valid for small R. Hence $\ln(\text{CurrentSalary}/\text{PreviousSalary}) = \ln(1 + R) \approx R = \beta \text{Performance}$ is linearly related to Performance. A regression of $\ln(\text{CurrentSalary}/\text{PreviousSalary})$ on Performance provides an estimate of β.

CHAPTER 2

Many excellent computer programs are available for plotting the data and for carrying out the regression calculations. Here we use S-Plus, R, Minitab, SAS, and SPSS. Most programs work the same and it is not difficult to switch from one program to the other. Most packages are spreadsheet programs. You enter the data into the various columns of a spreadsheet and use simple commands to carry out the operations. The results (fitted values, residuals, ...) can be stored in unused columns of the worksheet. Many options are available within all programs. You need to consult the on-line help for detailed discussion and examples.

The Minitab software is very easy to use. Minitab works like a spreadsheet program. We enter the data into columns of a spreadsheet and use the tabs: Stat > Regression > Regression. We specify the response variable and the explanatory (regressor) variables and execute the regression command. The output provides the estimates, standard errors, t-ratios and probability values. It displays the ANOVA table and the coefficient of determination. The output (residuals and fitted values) can be stored in unused columns of the worksheet.

A note on computing with R

R is a free software which is available through the internet; it can be downloaded from http://cran.us.r-project.org/. It is very similar to the commercial package S-Plus. R is a language and an environment for statistical computing and graphics. It can be used with Windows 95 or later versions, a variety of Unix and Linux platforms, and Apple Macintosh (OS versions later than 8.6).

The most convenient way to use R is at a graphics work station running a windowing system. We have used R on UNIX machines to solve several of the exercises, and the following discussion assumes this set-up. If you are running R under Windows, you will need to make some minor adjustments.

R issues the prompt " >" whenever it expects input commands. Let us assume that the UNIX shell prompt is %. You can start the R program with the command **%R**. Then R will return with a banner line, and R commands may be issued at this point. The command
> **>help.start()**

starts the HTML interface for on-line help, using the web browser that is available at your computer. You can use the mouse to explore features of the help facility. The command for quitting an R session is
> **>q()**

At this point you will be asked whether you want to save the data from your R session.

R has an extensive help facility. You can get information on any specific function –
for example the natural logarithm – by typing

>**help(log) or >?log**

R is case-sensitive, so x and X refer to different variables. R operates on named data
structures. Data can be entered at the terminal or can be read from an external file.
Entering the elements of a vector x – consisting of the four numbers 2, 4, 5, and 7 –
one uses the R command

>**x <- c(2,4,5,7) or >x = c(2,4,5,7)**

This is an assignment statement using the function c(). Notice that the assignment
operator "<-" (which is the same as the "=" operator) consists of the two characters <
("less than") and - ("minus") and points to the object receiving the value of the
expression. For simplicity we use "=".

For the exercises in this book we read the data from an external file (a text file in
UNIX). In exercise 2.6, for example, we have modified the file **hooker** so that the first
four lines are as follows:

Temp AP
210.8 29.211
210.2 28.559
208.4 27.972

The first line of the file specifies a name for each variable in the data frame. The
subsequent lines include the values for each variable. To read an entire data frame, we
use the command

>**hook = read.table("hooker",header=T)**

The filename **hooker** is in quotes; header =T indicates that the first line includes the
names of the variables. The commands

>**Temp = hook[,1]; >AP=hook[,2]**

define the first column of the matrix "hook" as Temp and the second column as AP.
The statement

>**LnAP = 100*log(AP)**

results in a transformation of the variable AP; log(AP) is the natural log of AP.

The function for fitting simple or multiple linear regression models is lm(). For
instance, a simple linear regression of Temp on LnAP can be fit by issuing the
command

>**hookfit = lm(Temp~LnAP)**

The output object from the lm() command, "hookfit", is a fitted model object.
Information about the fitted model can be extracted from this file. For example,

>**summary(hookfit)**

prints a comprehensive summary of the results of the regression analysis including the
estimated coefficients, their standard errors, t–values and p-values (see the solution to
exercise 2.6).

The command
>**anova(hookfit)**
supplies the analysis of variance (ANOVA) table. The command
>**plot(LnAP,Temp)**
plots Temp (the y-coordinate) against LnAP (the x-coordinate). A graphics window opens automatically. The fitted line can be superimposed on the scatter plot by issuing the command
>**abline(hookfit)**
The command
>**qqnorm(hookfit$residuals)**
leads to a normal probability plot of the residuals where "residuals" is in the fitted model object "hookfit".

Our discussion has focused on the free software package R. Note that the commands and the output of S-Plus are pretty much the same.

In subsequent chapters (Chapters 4 - 8) we consider multiple linear regression models. These models can be fit quite easily with R (and S-Plus). Suppose we have data in the vectors y, x1, x2 and x3. We can fit a multiple linear regression of y on x1, x2, and x3 by using the command
>**mregfit=lm(y~x1+x2+x3)**
Information about the model is in the fitted model object "mregfit". Note that an intercept term is included by default. One can restrict the intercept to be zero through
>**mulregfit=lm(y~x1+x2+x3-1)**

The above commands can be fine-tuned according to specific requirements. Many other commands are available to perform various statistical analyses and plots (such as residual analysis, leverages, Cook's D, various residual plots). This note is meant as a brief introduction to R. You should use the on-line help mentioned above to obtain more details.

2.1
(a) 95th percentile $= 10 + 3(1.645) = 14.93$; 99th percentile $= 10 + 3(2.326) = 16.98$
(b) $t(0.95;10) = 1.812$; $t(0.95;25) = 1.708$; $t(0.99;10) = 2.764$; $t(0.99;25) = 2.485$
(c) $\chi^2(0.95;1) = 3.84$; $\chi^2(0.95;4) = 9.49$; $\chi^2(0.95;10) = 18.31$
$\chi^2(0.99;1) = 6.63$; $\chi^2(0.99;4) = 13.28$; $\chi^2(0.99;10) = 23.21$
(d) $F(0.95;2,10) = 4.10$; $F(0.95;4,10) = 3.48$; $F(0.99;2,10) = 7.56$;
$F(0.99;4,10) = 5.99$

2.3 Correlation $= 0.816$; $R^2 = 0.867$; Estimated equation: $\hat{\mu} = 3 + 0.5x$

Same (linear regression) results for all four data sets. However, scatter plots in Figure 4.10 of the text show that linear regression is only appropriate for first data set. The correlation coefficients and the least squares estimates can be obtained by computer programs such as S-Plus, R, Minitab, SPSS, Minitab and others.

2.6
(a) Scatter plot (not shown here) indicates that a linear model is not appropriate. A quadratic component or a transformation are needed.
(b) Scatter plot confirms linear relationship between y = TEMP and x = 100ln(AP).
(c) R (or S-Plus) output from the function 'lm':

```
               Value   Std. Error   t value   Pr(>|t|)
(Intercept)   49.2684     1.1990     41.0925    0.0000
100ln(AP)      0.4782     0.0040    119.0838    0.0000

Residual standard error: s = 0.4016 with 29 degrees of freedom
Multiple R-Squared: 0.998
F-statistic: 14,180 with 1 and 29 degrees of freedom; the p-value is 0
```

(c) Estimated equation: $\hat{\mu} = 49.268 + 0.478\ln(AP)$; $R^2 = 0.998$; $s = \sqrt{MSE} = 0.402$.
   The model is appropriate since there is small random scatter around the fitted line;

(d) (i) $\hat{\beta}_1 = 0.4782$ and $s.e.(\hat{\beta}_1) = 0.0040$. Since $t(0.975;29) = 2.045$, a 95% confidence interval for β_1: $0.4782 - 2.045(0.0040)$, $0.4782 + 2.045(0.0040)$, or $(0.470, 0.486)$

   (ii) $\hat{\mu} = 49.268 + 0.478(100\ln(25)) = 203.195$;

$$s.e.(\hat{\mu}) = \sqrt{s^2\left[\frac{1}{n} + \frac{(x_0 - \bar{x})^2}{S_{xx}}\right]} = \sqrt{(0.402)^2/31 + (0.0040)^2(321.888 - 298.041)^2} = 0.1196$$

   95% confidence interval:
   $[203.195 - 2.045(0.1196), 203.195 + 2.045(0.1196)]$, or $(202.950, 203.440)$

(e) Estimates and standard errors of β_0 and β_1 change by factor of 5/9.

2.8 Minitab output:
```
The regression equation is
Revenue = 32 + 0.263 Cars

Predictor        Coef     SE Coef          T         P
Constant         31.9       185.2       0.17     0.867
Cars          0.26251     0.03930       6.68     0.000

S = 264.0      R-Sq = 84.8%      R-Sq(adj) = 82.9%
```

```
Analysis of Variance

Source              DF          SS          MS          F          P
Regression           1     3109923     3109923      44.62      0.000
Residual Error       8      557529       69691
Total                9     3667452
```

(a) Estimated equation: $\hat{\mu} = 31.9 + 0.2625x$; t-ratio($\hat{\beta}_1$) = 0.2625/0.0393 = 6.68;

p-value = 0.0002; number of cars sold is a significant predictor variable.

(b) 95% confidence interval for β_1: $0.2625 \pm (2.306)(0.0393)$ or (0.172, 0.353)

(c) $R^2 = 0.848$

(d) Standard deviation of y after factoring in x is $s = \sqrt{MSE} = 264.0$; standard deviation of y (without factoring x) is 638.3531.

(e) $\hat{\mu}(x = 1187) = 343.5$

2.10

(a) Prediction at weight 2000 is 0.5598 + (0.001024)(2000) = 2.6078. Since n is large and the estimation error can be ignored, s.e(prediction error) = s = $\sqrt{0.066}$ = 0.2569. Thus, an approximate 95% prediction interval is 2.6078 $\pm$ (1.96)(0.2569), or (2.104, 3.111). Note that 1.96 is from the standard normal table.

(b) The prediction at weight 1500 is 0.5598 + (0.001024)(1500) = 2.0958. Thus, an approximate 95% prediction interval is 2.09 $\pm$ (1.96)(0.2569) = (1.592, 2.599)

2.11

$$\frac{1}{R^2} = \frac{SST}{SSR} = \frac{SSR + SSE}{SSR} = 1 + \frac{SSE}{SSR} = 1 + \frac{n-p-1}{p}\frac{1}{F}$$

$$\text{Hence, } R^2 = \left[1 + \frac{n-p-1}{pF}\right]^{-1}.$$

2.13

(a) Estimated equation: $\hat{\mu} = 0.520x$; $s^2 = 46.2/16 = 2.89$;

$\hat{\beta}_1 = 0.520$; s.e.($\hat{\beta}_1$) = 0.0132 ; 95% confidence interval: (0.492, 0.548)

(b) Estimated equation: $\hat{\mu} = 0.725 + 0.498x$; $\hat{\beta}_0 = 0.725$; s.e.($\hat{\beta}_0$) = 1.549 ;

$\hat{\beta}_0$ / s.e.($\hat{\beta}_0$) = 0.725/1.549 = 0.47 ; p-value = 0.65; conclude $\beta_0 = 0$

2.15 R output:
```
Residual Standard Error = 4.5629
R-Square = 0.6767
F-statistic (df=1, 5) = 10.4657
p-value = 0.0231

            Estimate   Std.Err   t-value   Pr(>|t|)
Intercept   68.4459    12.9270   5.2948    0.0032
x           -0.4104    0.1268    -3.2351   0.0231

ANOVA
Source            DF       SS       MS      F       P
Regression        1      217.90   217.90  10.47   0.023
Residual Error    5      104.10    20.82
Total             6      322.00
```

(a) Estimated equation: $\hat{\mu} = 68.45 - 0.41x$; $R^2 = 0.677$; $s = 4.563$.

F-statistic = 10.47; p-value = 0.023; reject $\beta_1 = 0$

(b) s.e.$(\hat{\beta}_0) = 12.93$; $\hat{\beta}_0 / $s.e.$(\hat{\beta}_0) = 68.45/12.93 = 5.29$; p-value = 0.003

s.e.$(\hat{\beta}_1) = 0.127$; $\hat{\beta}_1 / $s.e.$(\hat{\beta}_1) = -0.41/0.127 = -3.23$; p-value = 0.023;

reject $\beta_0 = 0$ and $\beta_1 = 0$ at the 5 percent significance level.

99% confidence interval for β_1 : (-0.92, 0.11).

(c) $\hat{\mu}(x = 100) = 27.41$; s.e.$(\hat{\mu}(x = 100)) = 1.73$;

95% confidence interval: (22.97,31.86).

(d) $\hat{\mu}(x = 84) = 33.98$; s.e.$(\hat{\mu}(x = 84)) = 2.76$;

95% confidence interval: (26.88, 41.07).

Note that $\bar{x} = 101$ and $s.e.(\hat{\mu}_0)$ is smallest when $x_0 = \bar{x}$. As x_0 moves away from

$\bar{x}$, $s.e.(\hat{\mu}_0)$ becomes larger and the corresponding confidence interval becomes wider.

2.17

(a) The scatter plot shows that length (y) increases with increasing width (x).

```
Residual Standard Error = 4.295
R-Square = 0.9555
F-statistic (df=1, 8) = 171.7821
p-value = 0

              Estimate   Std.Error   t-value    Pr(>|t|)
Intercept     -46.4359   13.4161     -3.4612    0.0086
Width (x)       1.7924    0.1368     13.1066    0.0000
```

(b) Estimated equation: $\hat{\mu} = -46.44 + 1.792x$;

95% confidence interval for β_0 : (-77.37, -15.50);

95% confidence interval for β_1 : (1.48, 2.11).

(c) Good fit; $R^2 = 0.956$

(d) $\hat{\mu}(x = 100) = 132.8$; 95% prediction interval: (122.39,143.22)

(e) Strong linear relationship

2.18

(a) The plot of SBP against age indicates that there is a linear relationship between SBP and age.

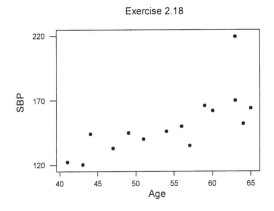

Exercise 2.18

(b) Estimated equation: $\hat{\mu} = 33.31 + 2.168x$;

(c) Analysis of variance

Source	DF	SS	MS	F	P
Regression	1	4361.5	4361.5	14.58	0.002
Residual Error	13	3889.4	299.2		
Total	14	8250.9			

(d) F = 14.58; p-value = 0.002; reject $\beta_1 = 0$

(e) s.e.$(\hat{\beta_1}) = 0.568$; $\hat{\beta_1} / $s.e.$(\hat{\beta_1}) = 2.168/0.568 = 3.82$; same p-value = 0.002; reject $\beta_1 = 0$

(f) Individual with $x = 63$ and $y = 220$ unusual. Estimates and standard errors change; R^2 increases. See R output shown below.

```
Residual Standard Error = 8.9007
R-Square = 0.7019
F-statistic (df=1, 12) = 28.2562
p-value=2e-04
```

13

```
              Estimate   Std.Error   t-value    Pr(>|t|)
Intercept     58.9876    16.6075     3.5519     4e-03
Weight         1.6244     0.3056     5.3157     2e-04


ANOVA

Source            DF        SS       MS       F       P
Regression         1     2238.5   2238.5   28.26   0.000
Residual Error    12      950.7     79.2
Total             13     3189.2
```

2.19 R Output:

```
Residual Standard Error = 0.1512
R-Square = 0.9496
F-statistic (df=1, 4) = 75.4083
p-value = 0.001

              Estimate   Std.Error   t-value    Pr(>|t|)
Intercept       3.7073    0.0955    38.8347      0.000
Mol.weight     -0.0123    0.0014    -8.6838      0.001
```

(a) Estimated equation: $\hat{\mu} = 3.707 - 0.0123x$; $R^2 = 0.950$

(b) F-statistic = 75.41; p-value = 0.001; reject $\beta_1 = 0$ at the 0.01 significance level. Significant linear relationship.

(c) Response is average of 3 observations. Use of individual values would improve the sensitivity of the analysis.

(d) No; molecular weight 200 far outside the region of experimentation; one does not know whether the linear relationship will continue to hold.

2.21 Plot of the chemical test against the magnetic test (not shown) indicates a linear relationship. Results of fitting a linear regression model are given below (R output):

```
Residual Standard Error = 3.4636
R-Square = 0.5372
F-statistic (df=1, 51) = 59.2056
p-value = 0

              Estimate   Std.Err   t-value   Pr(>|t|)
Intercept       8.9565    1.6523    5.4205      0
Mag Test        0.5866    0.0762    7.6945      0
```

Estimated equation: $\hat{\mu} = 8.957 + 0.587x$; $R^2 = 0.537$; F = 59.21; reject $\beta_1 = 0$
Significant linear relationship between the tests. However, variability large and predictive power low.

2.22 Plot of y (memory retention) against x (time) shows a nonlinear (exponentially decaying) pattern. Graphs of ln(y) against x and ln(y) against ln(x) show similar patterns. Plot of y against ln(x) shows a linear pattern.
Estimated equation: $\hat{\mu} = 0.846 - 0.079\ln(x)$; $R^2 = 0.990$; good model

2.27
(a) Response y = takeup(kg). Scatter plot indicates a linear relationship.
    y = Takeup(kg): $\hat{\mu} = -9.896 + 0.0753x$; $R^2 = 0.986$; F = 1,530.3; reject $\beta_1 = 0$

R Output

```
Residual Standard Error = 3.3945
R-Square = 0.9858
F-statistic (df=1, 22) = 1530.289     p-value = 0

              Estimate  Std.Error   t-value    Pr(>|t|)
Intercept      -9.8960    1.6887    -5.8602       0
x               0.0753    0.0019    39.1189       0
```

(b) Response y = takeup(kg). Scatter plot indicates a linear relationship.
    y = Takeup(%): $\hat{\mu} = 4.737 + 0.00162x$; $R^2 = 0.703$; F = 52.07; reject $\beta_1 = 0$

R Output

```
Residual Standard Error = 0.3952
R-Square = 0.703
F-statistic (df=1, 22) = 52.068
p-value = 0

              Estimate  Std.Error   t-value    Pr(>|t|)
Intercept       4.7372    0.1966    24.0973       0
x               0.0016    0.0002     7.2158       0
```

Both models fit well. However, the first one seems to be better (larger R^2).

CHAPTER 3

3.1

(a) $A = \begin{bmatrix} 2 & 0 & 1 \\ 3 & 2 & 2 \\ 2 & 1 & 4 \end{bmatrix}$; $A' = \begin{bmatrix} 2 & 3 & 2 \\ 0 & 2 & 1 \\ 1 & 2 & 4 \end{bmatrix}$

(b) $A'A = \begin{bmatrix} 2 & 3 & 2 \\ 0 & 2 & 1 \\ 1 & 2 & 4 \end{bmatrix} \begin{bmatrix} 2 & 0 & 1 \\ 3 & 2 & 2 \\ 2 & 1 & 4 \end{bmatrix} = \begin{bmatrix} 17 & 8 & 16 \\ 8 & 5 & 8 \\ 16 & 8 & 21 \end{bmatrix}$

(c) $\text{tr}(A) = 2 + 2 + 4 = 8$; $\text{tr}(A'A) = 17 + 5 + 21 = 43$

(d) $\det(A) = (2)(2)(4) + (3)(1)(1) + (0)(2)(2) - (2)(2)(1) - (1)(2)(2) - (3)(0)(4) = 11$

$\det(A'A) = (17)(5)(21) + (8)(8)(16) + (8)(8)(16) - (16)(5)(16) - (8)(8)(17) - (8)(8)(21)$
$= 121$

3.2

(a) $X'X = \begin{bmatrix} 4 & 0 & 0 \\ 0 & 4 & 0 \\ 0 & 0 & 4 \end{bmatrix}$; $(X'X)^{-1} = \begin{bmatrix} 1/4 & 0 & 0 \\ 0 & 1/4 & 0 \\ 0 & 0 & 1/4 \end{bmatrix}$; $X'y = \begin{bmatrix} 19 \\ 1 \\ 5 \end{bmatrix}$;

$(X'X)^{-1} X'y = \begin{bmatrix} 4.75 \\ 0.25 \\ 1.25 \end{bmatrix}$

(b) Diagonal matrices; the diagonal elements are the same.

3.5
(a) $\det(A) = (3)(4)(2) + (1)(1)(2) + (1)(1)(2) - (1)(1)(4) - (1)(1)(2) - (2)(2)(3) = 10$.

The inverse is given by $A^{-1} = \begin{bmatrix} 0.4 & 0 & -0.2 \\ 0 & 0.5 & -0.5 \\ -0.2 & -0.5 & 1.1 \end{bmatrix}$. Check that $AA^{-1} = A^{-1}A = I$.

You can use a computer program to determine the inverse and also to check your calculations.

(b) The three eigenvalues are the solutions to the cubic equation $\left| A - \lambda I \right| = 0$. They are given by 5.8951, 2.3973, and 0.7076. The corresponding eigenvectors are the columns of the matrix

$$P = \begin{bmatrix} -0.4317 & 0.8857 & 0.1706 \\ -0.7526 & -0.4579 & 0.4732 \\ -0.4973 & -0.0759 & -0.8643 \end{bmatrix}$$

(c) The spectral representation of the matrix A is given by

$$A = P\Lambda P' == \begin{bmatrix} -0.4317 & 0.8857 & 0.1706 \\ -0.7526 & -0.4579 & 0.4732 \\ -0.4973 & -0.0759 & -0.8643 \end{bmatrix} \begin{bmatrix} 5.8951 & 0 & 0 \\ 0 & 2.3973 & 0 \\ 0 & 0 & 0.7076 \end{bmatrix} \begin{bmatrix} -0.4317 & 0.8857 & 0.1706 \\ -0.7526 & -0.4579 & 0.4732 \\ -0.4973 & -0.0759 & -0.8643 \end{bmatrix}'$$

(d) The eigenvalues of A are positive, hence the matrix A is positive definite. The matrix A can be a covariance matrix. The correlation matrix is given by

$$\begin{bmatrix} 3/3 & 1/\sqrt{(3)(4)} & 1/\sqrt{(3)(2)} \\ 1/\sqrt{(3)(4)} & 4/4 & 2/\sqrt{(4)(2)} \\ 1/\sqrt{(3)(2)} & 2/\sqrt{(4)(2)} & 2/2 \end{bmatrix} = \begin{bmatrix} 1 & 0.289 & 0.408 \\ 0.289 & 1 & 0.707 \\ 0.408 & 0.707 & 1 \end{bmatrix}.$$

3.7
(a) $\det(A) = (2)(2)(6) + (1)(3)(3) + (1)(3)(3) - (3)(2)(3) - (1)(1)(6) - (3)(3)(2) = 0.$

(b) The three eigenvalues are the solutions to the cubic equation $\left| A - \lambda I \right| = 0$. They are given by 9, 1, and 0. The corresponding eigenvectors are the columns of the matrix

$$P = \begin{bmatrix} 1/\sqrt{6} & 1/\sqrt{2} & -1/\sqrt{3} \\ 1/\sqrt{6} & -1/\sqrt{2} & -1/\sqrt{3} \\ 2/\sqrt{6} & 0 & 1/\sqrt{3} \end{bmatrix}$$

(c) The eigenvalues are nonnegative; hence the matrix A is semi-positive definite. The matrix A can be a covariance matrix. The correlation matrix is given by

$$\begin{bmatrix} 2/2 & 1/\sqrt{(2)(2)} & 3/\sqrt{(2)(6)} \\ 1/\sqrt{(2)(2)} & 2/2 & 3/\sqrt{(2)(6)} \\ 3/\sqrt{(2)(6)} & 3/\sqrt{(2)(6)} & 6/6 \end{bmatrix} = \begin{bmatrix} 1 & 0.5 & 0.866 \\ 0.5 & 1 & 0.866 \\ 0.866 & 0.866 & 1 \end{bmatrix}$$

One eigenvalue is zero; hence there is a deterministic relationship among the three variables. The eigenvector corresponding to the eigenvalue 0 indicates the deterministic relationship. The linear combination $-y_1 - y_2 + y_3$ has variance zero.

3.8

(a) $AB = \begin{bmatrix} 1 & 4 & 2 \\ 3 & 1 & 2 \end{bmatrix} \begin{bmatrix} 4 & 1 \\ 2 & 2 \\ 2 & 4 \end{bmatrix} = \begin{bmatrix} 16 & 17 \\ 18 & 13 \end{bmatrix}$

(b) $BA = \begin{bmatrix} 4 & 1 \\ 2 & 2 \\ 2 & 4 \end{bmatrix} \begin{bmatrix} 1 & 4 & 2 \\ 3 & 1 & 2 \end{bmatrix} = \begin{bmatrix} 7 & 17 & 10 \\ 8 & 10 & 8 \\ 14 & 12 & 12 \end{bmatrix}$

3.11

(a) The distribution of $(y_1, y_2)'$ is bivariate normal with mean vector $(2,6)'$ and covariance matrix $\begin{bmatrix} 1 & 0 \\ 0 & 2 \end{bmatrix}$.

(b) The conditional distribution of $(y_1, y_2)'$, given that $y_3 = 5$, is bivariate normal with mean vector $\begin{bmatrix} 2 \\ 6 \end{bmatrix} + (1/3) \begin{bmatrix} 1 \\ -1 \end{bmatrix}(y_3 - 4) = \begin{bmatrix} (2/3) + (1/3)y_3 \\ (22/3) - (1/3)y_3 \end{bmatrix} = \begin{bmatrix} 7/3 \\ 17/3 \end{bmatrix}$ and covariance matrix $\begin{bmatrix} 1 & 0 \\ 0 & 2 \end{bmatrix} - (1/3) \begin{bmatrix} 1 \\ -1 \end{bmatrix} \begin{bmatrix} 1 & -1 \end{bmatrix} = \begin{bmatrix} 2/3 & 1/3 \\ 1/3 & 5/3 \end{bmatrix}$.

3.17 The quadratic form can be written as $y'Ay$ where the 3 x 3 symmetric matrix A is given as

$$A = \begin{bmatrix} 1 & 0 & 0 \\ 0 & 0.5 & 0.5 \\ 0 & 0.5 & 0.5 \end{bmatrix} .$$

The determinant of this matrix is 0. The rank of the matrix A is 2, as we can find a 2x2 submatrix with a nonzero determinant. Furthermore, the matrix A is idempotent; $AA = A$. Hence the distribution of the (normalized) quadratic form $\left(y_1^2 + 0.5y_2^2 + 0.5y_3^2 + y_2 y_3\right)/\sigma^2$ follows a chi-square distribution with 2 degrees of freedom.

CHAPTER 4

4.1

$$X'X = \begin{bmatrix} 10 & 55 \\ 55 & 385 \end{bmatrix}; \; (X'X)^{-1} = \begin{bmatrix} 0.4667 & -0.0667 \\ -0.0667 & 0.0121 \end{bmatrix};$$

$$V(\hat{\beta}) = \sigma^2 \begin{bmatrix} 0.4667 & -0.0667 \\ -0.0667 & 0.0121 \end{bmatrix}$$

$$V(\hat{\beta}_0) = (0.4667)\sigma^2; \; V(\hat{\beta}_1) = (0.0121)\sigma^2$$

4.5

(a) $V(\hat{\beta}_1) = 18$

(b) $Cov(\hat{\beta}_1, \hat{\beta}_3) = 1.2$

(c) $Corr(\hat{\beta}_1, \hat{\beta}_3) = 0.0943$

(d) $V(\hat{\beta}_1 - \hat{\beta}_3) = V(\hat{\beta}_1) + V(\hat{\beta}_3) - 2Cov(\hat{\beta}_1, \hat{\beta}_3) = 24.6$

4.7

(a) $R^2 = 0.9324$

(b) F-statistic = 110.35; p-value = 0.000; reject $\beta_1 = \beta_2 = \beta_3 = 0$

(c) 95% confidence interval for β_{taxes}: (0.074, 0.306); reject $\beta_{taxes} = 0$; cannot simplify model

   95% confidence Interval for β_{baths}: (-16.83, 180.57); can not reject $\beta_{baths} = 0$; can simplify model by dropping "baths"

4.10

(a) Estimated equation: $\hat{\mu} = 3.453 + 0.496x_1 + 0.0092x_2$; $s^2 = 4.7403$;

   s.e.$(\hat{\beta}_0) = 2.431$, s.e.$(\hat{\beta}_1) = 0.00605$, s.e.$(\hat{\beta}_2) = 0.00097$

(b) $t(\hat{\beta}_1) = 0.496 / 0.00605 = 81.89$; p-value (2-sided) = $2\,P(t(12) > 81.89) = 0.000$, which is very small. We reject the null hypothesis $\beta_1 = 0$.

   $t(\hat{\beta}_2) = 0.009191 / 0.00097 = 9.49$; p-value (2-sided) = 0.000, which is very small. We reject the null hypothesis $\beta_2 = 0$.

   Neither of the two explanatory variables can be omitted from the model.

4.12 The output from R software, using the function
lm(formula = usage ~ TEMP + PROD + DAYS + PAYR + HOUR) is given below:

```
Coefficients:
            Estimate   Std. Error   t value   Pr (>|t|)
(Intercept) 39.437054  12.110986    3.256     0.00765
TEMP         0.084067   0.060469    1.390     0.19194
PROD         0.001876   0.000607    3.091     0.01027
DAYS         0.131704   0.289800    0.454     0.65833
PAYS        -0.215677   0.098810   -2.183     0.05162
HOUR        -0.014475   0.030052   -0.482     0.63949
```

```
Residual standard error: 3.213 on 11 degrees of freedom
Multiple R-Squared: 0.6446, Adjusted R-squared: 0.4831
F-statistic: 3.991 on 5 and 11 DF, p-value: 0.02607
```

$R^2 = 0.6446$, and the regression model is significant at 2.6% level. The output indicates that PROD is significant at the 1% level, even if other variables are present in the model. PAYS is also marginally significant (p-value = 0.051). All other variables are not significant when added last to the model. The model can be simplified

(b) In order to test $\beta_1 = \beta_3 = \beta_5 = 0$, we need to fit a reduced model that includes just x_2 and x_4. The R output for the reduced model with lm(formula = USAGE ~ PROD + PAYR) is listed below

```
Coefficients:
            Estimate   Std. Error t value  Pr (>|t|)
(Intercept) 46.0177241 10.1085905   4.552  0.000452
PROD         0.0020353  0.0005587   3.643  0.002663
PAYR        -0.2157919  0.0895867  -2.409  0.030356
```

```
Residual standard error: 3.117 on 14 degrees of freedom
Multiple R-Squared: 0.5743, Adjusted R-squared: 0.5135
F-statistic: 9.442 on 2 and 14 DF, p-value: 0.002535
```

The additional sum of squares = ResidualSS (reduced model) – ResidualSS (full model) = SSR(full model) – SSR(reduced model) = 205.956 –183.48 and $F = [(205.956 –183.48)/2]/(3.213)^2 = 1.09$; p-value = $P(F(2,11) > 1.09) = 0.37$; we can not reject $\beta_1 = \beta_3 = \beta_5 = 0$.

(c) We prefer the reduced model $\hat{\mu} = 46.02 + 0.00204\text{PROD} - 0.216\text{PAYR}$; $R^2 = 0.574$ (only slightly smaller than the R^2 of the full model = 0.6446).

(d) Production has the smallest p-value.

(e) Water usage as linear function of PROD and PAYR. For fixed value of PAYR, each unit increase in production increases water use by 0.0020353 (gallons/100). Similarly, for a fixed value of PROD, a unit increase in PAYR decreases water usage by 0.2157919 (gallons/100).

4.14

(a)

$$X'X = \begin{bmatrix} 13.00 & 59.43 & 81.82 & 115.40 \\ 59.43 & 394.73 & 360.66 & 522.08 \\ 81.82 & 360.66 & 576.73 & 728.31 \\ 115.40 & 522.08 & 728.31 & 1035.96 \end{bmatrix}$$

$$(X'X)^{-1} = \begin{bmatrix} 8.06479464 & -0.082592705 & -0.094195115 & -0.790526876 \\ -0.08259271 & 0.008479816 & 0.001716687 & 0.003720020 \\ -0.09419511 & 0.001716687 & 0.016629424 & -0.002063308 \\ -0.79052688 & 0.003720020 & -0.002063308 & 0.088601286 \end{bmatrix}$$

$$X'y = \begin{bmatrix} 377.700 \\ 1877.911 \\ 2247.285 \\ 3339.300 \end{bmatrix}$$

(c) Estimated equation: $\hat{\mu} = 39.482 + 1.0092x_1 - 1.873x_2 - 0.367x_3$

(d) (i) (22.802, 25.653); 90% confidence interval for the mean value of y when $x_1 = 3$, $x_2 = 8$ and $x_2 = 9$ can be obtained with the software R directly using the function "predict".

Mean value	Lower limit	Upper limit
24.22764	22.80225	25.65302

There is also an option in Minitab.

  (ii) (20.109, 28.346); 90% prediction interval for an individual value of y when $x_1 = 3$, $x_2 = 8$ and $x_2 = 9$ can also be obtained from the software R directly using the function "predict".

(e) F-statistic = 30.08; p-value = 0.000; reject $\beta_1 = \beta_2 = \beta_3 = 0$.

4.16

$$(X'X)^{-1} = \begin{bmatrix} 9.61093203 & 0.008587789 & -0.27914754 & -0.04452169 \\ 0.00858779 & 0.509964070 & -0.25886359 & 0.00077654 \\ -0.27914754 & -0.25886359 & 0.13949996 & 0.00073956 \\ -0.04452169 & 0.00077654 & 0.00073956 & 0.00036978 \end{bmatrix}$$

Correction Factor = $45^2/9 = 225$

SST = $y'y - CF = 285 - 225 = 60$

SSR = $\hat{\beta}X'y - CF = 282.9725 - 225 = 57.9725$

SSE = SST - SSR $= 60 - 57.9725 = 2.0275$

ANOVA table:

```
Source       DF     SS       MS       F       P
Regression    3   57.9725  19.3242  47.66   2.129815e-05
Residual      5    2.0275   0.4055
Total         8   60.0000
```

F-statistic = 47.66; reject $\beta_1 = \beta_2 = \beta_3 = 0$

(b) Estimated equation: $\hat{\mu} = -1.16346 + 0.13527x_1 + 0.01995x_2 + 0.12195x_3$;

$s^2 = 0.4055$;

s.e.($\hat{\beta}_0$) = 1.974; s.e.($\hat{\beta}_1$) = 0.45474; s.e.($\hat{\beta}_2$) = 0.23784; s.e.($\hat{\beta}_3$) = 0.01225

$t(\hat{\beta}_1) = 0.295$; p-value = 0.78; can not reject $\beta_1 = 0$

$t(\hat{\beta}_2) = 0.084$; p-value = 0.94; can not reject $\beta_2 = 0$

$t(\hat{\beta}_3) = 9.955$; p-value = 0.000; reject $\beta_3 = 0$

4.20

(a) $\hat{\beta}^{WLS} = \sum y_i / \sum x_i$; $V(\hat{\beta}^{WLS}) = \sigma^2 / \sum x_i$

(b) $\hat{\beta}^{WLS} = 30/2 = 15$; $V(\hat{\beta}^{WLS}) = \sigma^2 / 150$

4.22

(a) Linear model not appropriate.

(b) Fitted equation:

TensileStrength = -6.674 + 11.764 Hardwood - 0.635(Hardwood)^2

Source	DF	SS	MS	F	P
Regression	2	3104.2	1552.1	79.43	0.000
Residual Error	16	312.6	19.5		
Total	18	3416.9			

Model adequate; quadratic term needed; increases R^2 from 0.305 to 0.909.
95% confidence interval for mean response when hardwood 6 percent: (38.14, 44.00)
Prediction intervals are for individual observations while confidence intervals are for the mean value. Confidence intervals are shorter than the corresponding prediction intervals. 95% prediction interval for tensile strength for a batch of paper with 6 percent hardwood concentration: (31.25, 50.88).
The maximum hardwood concentration in the data set used to fit the model is 7 percent, which is very low compared to 20 percent. It is not advisable to use the fitted model to predict the mean tensile strength of paper for 20 percent hardwood concentration.

4.23

Quadratic model. Estimated equation: $\hat{\mu} = 82.385 - 38.310x + 4.703x^2$

Regression significant; adequate fit.
Stars with ln(surface temperature) < 4 appear different and should be investigated separately. Without these stars, a linear model is appropriate.

Predictor	Coef	SE Coef	T	P
Constant	82.385	9.581	8.60	0.000
x	-38.310	4.790	-8.00	0.000
x2	4.7025	0.5939	7.92	0.000

S = 0.3667   R-Sq = 60.6%   R-Sq(adj) = 58.8%

Analysis of Variance

Source	DF	SS	MS	F	P
Regression	2	9.0945	4.5472	33.82	0.000
Residual Error	44	5.9165	0.1345		
Total	46	15.01			

CHAPTER 5

5.2 (a) $ 3,000; (b) $ 900

5.4 $VIF_1 = 1/(1-R_1^2) = 2.5$; $VIF_2 = 1/(1-R_2^2) = 5$; $VIF_3 = 1/(1-R_3^2) = 10$;
evidence of multicollinearity since variance inflation factors are large (10 or larger).

5.6 Define two indicator variables x_1 and x_2 such that $x_1 = 0$ and $x_2 = 0$ represent the group Sparrow, $x_1 = 1$, $x_2 = 0$ represent Robin, and $x_1 = 0$ and $x_2 = 1$ represent Wren. Then the model can be expressed as $E(y) = \beta_0 + \beta_1 x_1 + \beta_2 x_2$ in which
$\beta_1 = \mu(\text{Robin}) - \mu(\text{Sparrow})$ and $\beta_2 = \mu(\text{Wren}) - \mu(\text{Sparrow})$.

```
Analysis of Variance
                                Sum of          Mean
   Source              DF      Squares        Square   F Value   Pr > F
   Model                2     31.11193      15.55596     22.33   <.0001
   Error               42     29.26052       0.69668
   Corrected Total     44     60.37244
```

F-statistic = 22.33 tests whether there are differences among the three group means; p-value < 0.0001; reject H_0: $\mu_1 = \mu_2 = \mu_3$ (or $\beta_1 = \beta_2 = 0$)

5.8
(a) Expected difference in systolic blood pressure for females versus males who drink the same number of cups of coffee, excercise the same, and are of the same age
(b) Represents variation due to measurement error and omitted factors
(c) Association, but not causation
(d) Represents interaction between gender and coffee consumption

5.9

(a) $E(y_t) = \begin{cases} \beta_0 + \beta_1 t, & t = 1,2,...,7 \\ \beta_2 + \beta_3 t, & t = 8,9,...,14 \end{cases}$

Intersecting lines at t = 8: $\beta_2 = \beta_0 + 8(\beta_1 - \beta_3)$, and

$E(y_t) = \begin{cases} \beta_0 + \beta_1 t, & t = 1,2,...,7 \\ \beta_0 + \beta_1 8 + \beta_3 (t-8), & t = 8,9,...,14 \end{cases}$

In matrix form, $E(y) = X\beta$ where

$$X = \begin{bmatrix} 1 & 1 & 0 \\ 1 & 2 & 0 \\ . & . & . \\ 1 & 7 & 0 \\ 1 & 8 & 0 \\ 1 & 8 & 1 \\ . & . & . \\ 1 & 8 & 6 \end{bmatrix} \quad \text{and} \quad \boldsymbol{\beta} = \begin{bmatrix} \beta_0 \\ \beta_1 \\ \beta_3 \end{bmatrix}$$

(b) $E(y_t) = \beta_0 + \beta_1 t$, $t = 1,2,...,14$

(c) $F = 55.95$; p-value $= P(F(1,11) > 55.95) = 0.0000$; model in (a) is preferable.

5.12

```
Analysis of Variance
                        Sum of        Mean
Source           DF     Squares      Squares   F Value   Pr > F
Model             4     39.37694     9.84423    14.07   <.0001
Error            25     17.49506     0.69980
Corrected Total  29     56.87200

                     Parameter     Standard
Variable     DF      Estimate        Error     t Value   Pr > |t|
Intercept     1      -0.91221       0.87548     -1.04     0.3074
x1            1       0.16073       0.06617      2.43     0.0227
x2            1       0.21978       0.03406      6.45     <.0001
x3            1       0.01123       0.00497      2.26     0.0330
x4            1       0.10197       0.05874      1.74     0.0948
```

(b) $\hat{\mu} = -0.9122 + 0.1607x_1 + 0.2198x_2 + 0.0112x_3 + 0.1020x_4$; $R^2 = 0.692$; s = 0.8365;

   (i) $t(\hat{\beta_1}) = 2.43$; p-value $= 0.023$; reject $\beta_1 = 0$

   (ii) $F = (5.45747/2)/(0.69980) = 3.90$ (use of additional SS); p-value $= 0.034$; reject the null hypothesis $\beta_3 = \beta_4 = 0$

   (iii) $F=14.07$; p-value $<.0001$; reject hypothesis $\beta_1 = \beta_2 = \beta_3 = \beta_4 = 0$.

(c)
$\hat{\mu} = -1.462 + 0.1536x_1 + 0.3221x_2 + 0.0166x_3 + 0.0571x_4 - 0.00087x_2 x_3 + 0.00599x_2 x_4$

   $H_0 : \beta_5 = \beta_6 = 0$: F $= 0.40$; p-value $= 0.67$; interactions not important.

(d) (i) Since all coefficients are positive: Lower wrinkle resistance for lower x_1, x_2, x_3, and x_4.

(ii) Increased wrinkle resistance for higher x_1, x_2, x_3, and x_4.

(e) It is difficult to generalize the conclusions from this study since the values of x_1, x_2, x_3, and x_4 were not controlled. One suggestion for improvement is to conduct an experiment in which the values of x_1, x_2, x_3, and x_4 are controlled and the resulting response y measured.

5.13

(b) $z = 0$ (protein-rich); $z = 1$ (protein-poor): $\hat{\mu} = 50.324 + 16.009x + 0.918z - 7.329xz$

H_0: $\beta_2 = \beta_3 = 0$. Test whether the linear relationship between height (y) and age (x) is the same for the two diets. Additional SS = ResidualSS (reduced model) – ResidualSS (full model) = 1120.22, and $F = (1120.22/2)/(5.22290) = 107.24$; p-value < 0.0001; reject $\beta_2 = \beta_3 = 0$; linear relationships between height and age not the same for the two diets.

5.15 Weight (x_1); $x_2 = 0$ (type A engine); $x_2 = 1$ (type B engine);

(a) $\mu = \beta_0 + \beta_1 x_1 + \beta_2 x_2$; (b) $\mu = \beta_0 + \beta_1 x_1 + \beta_2 x_2 + \beta_3 x_1 x_2$

CHAPTER 6

6.2

(a) Linear model: $\hat{\mu} = 23.35 + 1.045x$; $R^2 = 0.955$; $s = 0.737$;

F(lack of fit) = 10.01; p-value = 0.002; lack of fit.

Source	d.f	S.S	M.S	F	Prob$\geq$F
Model	1	195.2428	195.2428	359.3	0.0001
Error	17	9.2382	0.5434		
Lack of Fit	9	8.4849	0.9427	10.01	<0.01
Pure Error	8	0.7533	0.0942		

(b) Quadratic model: $\hat{\mu} = 22.56 + 1.67x - 0.068x^2$; $R^2 = 0.988$; $s = 0.394$;

$t(\hat{\beta}_2) = -0.06796 / 0.01031 = -6.59$; reject $\beta_2 = 0$;

F(lack-of-fit) = 2.30; p-value = 0.13; no lack of fit.

Source	d.f	S.S	M.S	F	Prob>F
Model	2	201.9944	100.9972	649.86	0.0001
Error	16	2.4866	0.1554		
Lack of Fit	8	1.7333	0.2166	2.3	>.10
Pure Error	8	0.7533	0.0947		

6.7

(a) True. For a correct model, $Cov(e, \hat{\mu}) = O$, and a plot of the residuals e_i against the fitted values $\hat{\mu}_i$ should show no association. However, $Cov(e, y) = \sigma^2(I - H)$; the correlation makes the interpretation of the plot of e_i against y_i difficult.

(b) Not true. Outliers should be scrutinized, but not necessarily rejected.

(c) True

6.8 (a) 5; (b) 2; (c) 4; (d) 1

6.11 (a) No; (b) No; (c) No; (d) No; (e) True

6.12 A (Palm Beach); B (Broward); C (Dade); D (Pasco)

6.15 Scatter plots of y , ln(y) and 1/y against x point to a log transformation. The estimate of the transformation parameter in Box-Cox family is $\hat{\lambda} \approx 0$, indicating a logarithmic transformation of the response y.

Regression of ln(y) on x: $\hat{\mu} = 2.436 + 0.000567x$; $R^2 = 0.986$; s = 0.0845.

The first case is quite influential (x = 574; y = 21.9; Cook = 0.585).

Box -Cox transformation

λ	$s(\lambda)$	R^2
-1.00	11.270	0.922
-0.75	8.569	0.948
-0.50	6.331	0.969
-0.25	4.690	0.982
-0.10	4.165	0.985
0.001 (ln)	4.082	0.986
0.10	4.232	0.985
0.25	4.849	0.980
0.50	6.629	0.965
0.75	9.033	0.942
1.00	11.960	0.912

$s(\lambda)$ is the residual standard error and R^2 is the coefficient of determination in the

regression of $\dfrac{y^{\lambda} - 1}{\lambda (\bar{y}_g)^{\lambda-1}}$ on x.

6.21

Scatter plot of ln(y) against ln(x) shows a linear association with three outlying observations (brachiosaurus, diplodocus, and triceratops). Omitting these three cases and fitting the linear model to the reduced data set leads to an adequate fit.
Estimated equation: $\hat{\mu} = 2.15 + 0.752 \ln(x)$; $R^2 = 0.922$; s = 0.726. The two observations with the largest positive residuals and the largest Cook influence are human (stand. residual = 2.72; Cook = 0.174) and Rhesus monkey (stand. residual = 2.25; Cook =0.119).

6.22

Estimated equation: $\hat{\mu} = 74.319 - 2.089 \text{Conc} + 0.430 \text{Ratio} - 0.372 \text{Temp}$;
$R^2 = 0.939$; s = 0.74; F(lack of fit) = 7.44; p-value = 0.036; indication of lack of fit.

```
Analysis of Variance

Source             DF          SS          MS          F          P
Regression          3      92.304      30.768      56.17      0.000
Residual Error     11       6.026       0.548
  Lack of Fit       7       5.596       0.799       7.44      0.036
  Pure Error        4       0.430       0.108
Total              14      98.329
```

Run #2 (Conc = 1, Ratio = -1, Temp = -1; Yield = 73.9) influential, with large Cook's distance. This run should be investigated. Without this run, no lack of fit.

6.26

Linear model: $\hat{\mu} = 0.131 + 0.241x$, with $R^2 = 0.874$, is not appropriate.

Quadratic model: $\hat{\mu} = -1.16 + 0.723x - 0.0381x^2$, with $R^2 = 0.968$, is a possibility.

90% confidence interval: (1.972, 2.102).

Reciprocal transformation on x: $\hat{\mu} = 2.98 - 6.93(1/x)$, with $R^2 = 0.980$, is better.

90% confidence interval: (1.951, 2.026).

CHAPTER 7

7.1

(a) Backward elimination: Drop x_3 (step 1); drop x_4 (step 2); next candidate x_2 for elimination can not be dropped. Model with x_1 and x_2.

(b) Forward selection: Enter x_4 (step1); enter x_1 (step 2); enter x_2 (step 3); next candidate x_3 for selection can not be entered. Model with x_1, x_2, and x_4.

(c) Stepwise Regression: Steps 1, 2 and 3 of forward selection; x_4 can be dropped from the model containing x_1, x_2, and x_4; no reason to add x_3 to the model with x_1 and x_2. Model with x_1 and x_2.

(d) Model with x_1 and x_2: $C_p = 2.68$, close to desired value 3. Full model: $C_p = 5$. Prefer model with x_1 and x_2.

(e) x_2 and x_4 are highly correlated.

(f) $F = 68.6$; p-value less than 0.001; reject $\beta_1 = \beta_3 = 0$.

7.2

(a) C_p: Model with x_1 and x_2 ($C_p = 2.7$)

$R_?^2$: Model with x_1 and x_2, or model with x_1 and x_4. Small gain by going to more complicated models.

(b) Backward elimination ($\alpha_{drop} = 0.1$): Model with x_1 and x_2.

Forward selection ($\alpha_{enter} = 0.1$): Model with x_1, x_2, and x_4.

Stepwise regression ($\alpha_{drop} = \alpha_{enter} = 0.1$): Model with x_1 and x_2.

7.7

$\hat{\mu} = -5.0359 + 0.0671\text{AirFlow} + 0.1295\text{CoolTemp}$; $R^2 = 0.909$; $C_p = 2.9$.

Last case (AirFlow = 70; CoolTemp = 20; StackLoss = 1.5) is an influential observation and should be scrutinized. Without this case:

$\hat{\mu} = -5.1076 + 0.0863\text{AirFlow} + 0.0803\text{CoolTemp}$; $R^2 = 0.946$

7.8

Stepwise regression ($\alpha_{drop} = \alpha_{enter} = 0.15$):

$\hat{\mu} = -62.60 + 7.427\%\text{ASurf} + 6.828\%\text{ABase} - 5.2685\text{Run}$;

$R^2 = 0.724$; $R_{adj}^2 = 0.693$; $C_p = 1.3$.

Similar model: $\hat{\mu} = -23.00 + 5.975\%ASurf - 5.4058Run$;

$R^2 = 0.695$; $R^2_{adj} = 0.673$; $C_p = 1.9$.

Cases 13 and 15 with large Cook's influence. Second set of runs with considerably smaller change in rut depth.

CHAPTER 8

(a) The Minitab output of various regression models is given below. For each fitted model we list the estimated equation (with estimates, standard errors, and p-values), the coefficient of determination R^2, the root mean square error s, and the Durbin-Watson statistic. Minitab flags observations with unusually large standardized residuals ("R") and with unusually large leverage ("X"). The Lockerbie model is simplified by omitting insignificant variables.

Campbell (n = 13):

```
Incumbent Vote = 25.8 + 0.492 Sept Trial + 2.26 GDP Growth

Predictor        Coef      SE Coef          T        P
Constant       25.754        2.953       8.72    0.000
Sept Trial    0.49173      0.05716       8.60    0.000
GDP Growth     2.2571       0.4921       4.59    0.001

S = 1.827      R-Sq = 92.2%     R-Sq(adj) = 90.7%

Unusual Observations
Obs    Sept Tri   Incumbent       Fit     SE Fit   Residual    St Resid
  9       48.7      44.700    44.226      1.570      0.474        0.51 X

X denotes an observation whose X value gives it large influence.

Durbin-Watson statistic = 2.15
```

Abramowitz(n = 13):

```
Incumbent Vote = 45.1 - 4.69 Term + 0.179 Popularity + 2.14 GDP Growth

Predictor        Coef      SE Coef          T        P
Constant       45.059        2.865      15.73    0.000
Term           -4.691        1.337      -3.51    0.007
Popularity     0.17855      0.05567      3.21    0.011
GDP Growth     2.1389       0.6352       3.37    0.008

S = 1.984      R-Sq = 91.7%     R-Sq(adj) = 89.0%

Unusual Observations
Obs     Term    Incumbent       Fit     SE Fit   Residual    St Resid
 13     0.00      54.600    58.480      0.929     -3.880       -2.21R

R denotes an observation with a large standardized residual

Durbin-Watson statistic = 1.76
```

Holbrook (n = 13):

Incumbent Vote = 17.6 + 0.0998 PresPop + 0.296 PersFin - 4.00 Tenure

```
Predictor        Coef      SE Coef          T        P
Constant       17.606        3.865       4.56    0.001
PresPop       0.09982      0.04668       2.14    0.061
PersFin       0.29589      0.04112       7.20    0.000
Tenure         -3.995        1.002      -3.99    0.003
```

S = 1.505   R-Sq = 95.3%   R-Sq(adj) = 93.7%

Durbin-Watson statistic = 2.07

Lockerbie (n = 11):

The regression equation is
Incumbent Vote = 22.4 + 0.635 Inc1 - 0.184 Inc2 + 1.13 NextYearBetter
          - 1.45 Tenure

```
Predictor        Coef      SE Coef          T        P
Constant       22.351        7.231       3.09    0.021
Inc1           0.6352       0.5136       1.24    0.262
Inc2          -0.1836       0.4923      -0.37    0.722
NextYear       1.1251       0.2103       5.35    0.002
Tenure        -1.4488       0.2489      -5.82    0.001
```

S = 1.661   R-Sq = 95.4%   R-Sq(adj) = 92.3%

Durbin-Watson statistic = 1.17

The regression equation is
Incumbent Vote = 21.4 + 0.604 Inc1 + 1.13 NextYearBetter - 1.39 Tenure

```
Predictor        Coef      SE Coef          T        P
Constant       21.423        6.359       3.37    0.012
Inc1           0.6044       0.4747       1.27    0.244
NextYear       1.1340       0.1956       5.80    0.001
Tenure        -1.3894       0.1793      -7.75    0.000
```

S = 1.555   R-Sq = 95.3%   R-Sq(adj) = 93.2%

Durbin-Watson statistic = 1.32

The regression equation is
Incumbent Vote = 16.6 + 1.30 NextYearBetter - 1.37 Tenure

```
Predictor        Coef      SE Coef          T        P
Constant       16.646        5.329       3.12    0.014
NextYear       1.3029       0.1493       8.73    0.000
Tenure        -1.3726       0.1857      -7.39    0.000
```

```
S = 1.615        R-Sq = 94.2%    R-Sq(adj) = 92.7%

Durbin-Watson statistic = 1.26
```

(b) The sample sizes for estimating these models is extremely small ($n = 13$ and $n = 11$). Considering the extremely small sample sizes, we can not detect violations of the assumption of independent errors.

(c) The root mean square errors for most fitted models are in the range from 1.5 to 2 percentage points. They are similar to the ones in the Fair and Lewis-Beck/Tien models. The size of the root mean square error implies that the half widths of 95% prediction intervals are <u>at least</u> 3 - 4 percentage points. Incorporating the uncertainty from the estimation and considering that the sample size is very small makes the prediction intervals even wider. Furthermore, the predictions are "within-sample" predictions, which means that the case being predicted is part of the data that are used for estimation. Prediction errors for "out-of-sample" predictions (where the case being predicted is not part of the data used for the estimation) are usually larger; see (d).

(d) Leaving out case *i*, running the regression on the reduced data set, and predicting the response of the case that has been left out using the estimates from the reduced data set, leads to the PRESS residuals $e_{(i)}$ in equation (6.21) of Chapter 6. Equation (6.22) implies that the PRESS residuals can be calculated from the regular residuals and the leverages. That is,

$$e_{(i)} = y_{(i)} - \hat{y}_{(i)} = e_i / (1 - h_{ii})$$

For illustration we have calculated the residuals, leverages and PRESS residuals for the regression model considered by Campbell in the beginning of this exercise. The PRESS residuals are larger than the ordinary residuals. For example, the (out-of-sample) prediction error for 1996 is -3.76.

Year	Incumbent Vote	Sept Trial	GDP Growth	Residuals	Leverage	PRESS
1948	52.32	45.61	0.91	2.08441	0.126153	2.38533
1952	44.59	42.11	0.27	-2.48002	0.166349	-2.97488
1956	57.75	55.91	0.64	3.05900	0.093183	3.37334
1960	49.92	50.54	-0.26	-0.09906	0.134083	-0.11439
1964	61.34	69.15	0.81	-0.24520	0.361195	-0.38384
1968	49.60	41.89	1.63	-0.43144	0.280740	-0.59984
1972	61.79	62.89	1.73	1.20653	0.235919	1.57906
1976	48.95	40.00	1.17	0.88618	0.257420	1.19338
1980	44.70	48.72	-2.43	0.47371	0.738021	1.80821
1984	59.17	60.22	1.79	-0.23597	0.203862	-0.29640
1988	53.90	54.44	0.79	-0.40671	0.083538	-0.44379

1992	46.55	41.94	0.35	-0.61699	0.168430	-0.74195
1996	54.74	60.67	1.04	-3.19446	0.151107	-3.76308

(e) The four prediction models studied in this exercise are no better and no worse than the models by Fair and Lewis-Beck/Tien. While they give us some indication about the winner of presidential elections, their large uncertainty makes them only useful in the rather uninteresting situation when there is little doubt about the winner of the election.

8.2

Part 1(a): Modeling the height and the weight at referral (HeightR, WeightR) as a function of age at referral (AgeR)

Models with a linear component of Age provide an adequate representation of the relationships. Addition of Age**2 is not necessary. The models lead to an R-square of about 60 percent for height, and 45 percent for weight. Height at referral is easier to predict than weight. Birth weight is marginally significant (estimate 2.26, with p-value 0.064). Addition of birth weight to the regression of weight at referral on age at referral increases the R-square from 45.9 to 48.3 percent. Each extra pound at birth increases the weight at referral by 2.26 pounds. Average weight at referral is 73 pounds, with standard deviation 20 pounds.

```
Regression Analysis: HeightR versus AgeR, AgeR**2

The regression equation is
HeightR = 19.1 + 0.452 AgeR - 0.00120 AgeR**2

77 cases used 16 cases contain missing values

Predictor          Coef     SE Coef         T        P
Constant         19.095       9.434      2.02    0.047
AgeR             0.4523      0.1700      2.66    0.010
AgeR**2      -0.0012036   0.0007501     -1.60    0.113

S = 2.999       R-Sq = 60.4%     R-Sq(adj) = 59.3%
```

Note: Because of the multicollinearity between AgeR and AgeR**2, both regression coefficients are (partially) insignificant. However, this does not imply that both can be omitted from the model at the same time. The results of the model given below show that AgeR is significant if it is the only variable in the model.

```
Regression Analysis: HeightR versus AgeR

The regression equation is
HeightR = 33.9 + 0.181 AgeR
```

77 cases used 16 cases contain missing values

```
Predictor        Coef     SE Coef          T        P
Constant       33.912       1.949      17.40    0.000
AgeR          0.18088     0.01741      10.39    0.000

S = 3.030      R-Sq = 59.0%     R-Sq(adj) = 58.5%
```

Regression Analysis: WeightR versus AgeR, AgeR**2

The regression equation is
WeightR = - 0.9 + 0.656 AgeR + 0.00009 AgeR**2

80 cases used 13 cases contain missing values

```
Predictor        Coef     SE Coef          T        P
Constant        -0.94       46.45      -0.02    0.984
AgeR           0.6555      0.8387       0.78    0.437
AgeR**2      0.000094    0.003704       0.03    0.980

S = 15.09      R-Sq = 45.9%     R-Sq(adj) = 44.5%
```

Regression Analysis: WeightR versus AgeR

The regression equation is
WeightR = - 2.09 + 0.677 AgeR

80 cases used 13 cases contain missing values

```
Predictor        Coef     SE Coef          T        P
Constant       -2.090       9.341      -0.22    0.824
AgeR          0.67658     0.08321       8.13    0.000

S = 14.99      R-Sq = 45.9%     R-Sq(adj) = 45.2%
```

Regression Analysis: WeightR versus AgeR, BirthWeight

The regression equation is
WeightR = - 16.1 + 0.653 AgeR + 2.26 BirthWeight

80 cases used 13 cases contain missing values

```
Predictor        Coef     SE Coef          T        P
Constant       -16.15       11.85      -1.36    0.177
AgeR          0.65326     0.08282       7.89    0.000
BirthWeight     2.259       1.202       1.88    0.064

S = 14.75      R-Sq = 48.3%     R-Sq(adj) = 46.9%
```

<u>Part 1(b)</u>: Modeling the height and the weight at follow-up (HeightF, WeightF) as a function of age at follow-up (AgeF)

Similar conclusions as in 1(a). Models with a linear component of Age provide an adequate representation of the relationships. Addition of Age**2 is not needed. The models lead to an R-square of about 40 percent for both height and weight. Birth weight is significant (estimate 4.97 with p-value 0.01). Each extra pound at birth increases the weight at follow-up by 5 pounds. Average weight at follow-up is 124 pounds, with standard deviation 32 pounds.

Regression Analysis: HeightF versus AgeF, AgeF2**

```
The regression equation is
HeightF = 10.0 + 0.458 AgeF - 0.00080 AgeF**2
```

81 cases used 12 cases contain missing values

Predictor	Coef	SE Coef	T	P
Constant	10.02	34.71	0.29	0.774
AgeF	0.4581	0.3937	1.16	0.248
AgeF**2	-0.000795	0.001106	-0.72	0.474

S = 4.115   R-Sq = 41.8%   R-Sq(adj) = 40.3%

Regression Analysis: HeightF versus AgeF

```
The regression equation is
HeightF = 34.8 + 0.176 AgeF
```

81 cases used 12 cases contain missing values

Predictor	Coef	SE Coef	T	P
Constant	34.801	4.090	8.51	0.000
AgeF	0.17553	0.02347	7.48	0.000

S = 4.103   R-Sq = 41.4%   R-Sq(adj) = 40.7%

Regression Analysis: WeightF versus AgeF, AgeF2**

```
The regression equation is
WeightF = - 158 + 2.23 AgeF - 0.00339 AgeF**2
```

85 cases used 8 cases contain missing values

Predictor	Coef	SE Coef	T	P
Constant	-158.2	206.4	-0.77	0.445
AgeF	2.227	2.349	0.95	0.346
AgeF**2	-0.003387	0.006620	-0.51	0.610

S = 25.24   R-Sq = 39.9%   R-Sq(adj) = 38.5%

Regression Analysis: WeightF versus AgeF

The regression equation is
WeightF = - 53.4 + 1.03 AgeF

85 cases used 8 cases contain missing values

Predictor	Coef	SE Coef	T	P
Constant	-53.37	24.17	-2.21	0.030
AgeF	1.0269	0.1388	7.40	0.000

S = 25.13   R-Sq = 39.7%   R-Sq(adj) = 39.0%

Regression Analysis: WeightF versus AgeF, BirthWeight

The regression equation is
WeightF = - 82.0 + 0.982 AgeF + 4.97 BirthWeight

85 cases used 8 cases contain missing values

Predictor	Coef	SE Coef	T	P
Constant	-82.04	25.84	-3.18	0.002
AgeF	0.9815	0.1353	7.25	0.000
BirthWeight	4.967	1.910	2.60	0.011

S = 24.30   R-Sq = 44.3%   R-Sq(adj) = 43.0%

Part 1(c): Modeling the combined data: HeightCo, WeightCo and AgeCo.

Models with a linear component of AgeCo provide an adequate representation of the relationship between HeightCo and AgeCo. For weight, the addition of the quadratic component AgeCo**2 becomes necessary. The scatter plot of weight against age suggests that the variability increases with the level. The scatter plot of the logarithm of weight against age indicates that the variability is stabilized by this transformation. The residuals from the regression of ln(WeightCo) on AgeCo are unremarkable. No major lack of fit can be detected.

Regression Analysis: HeightCo versus AgeCo, AgeCo2**

The regression equation is
HeightCo = 31.3 + 0.221 AgeCo -0.000144 AgeCo**2

158 cases used 28 cases contain missing values

Predictor	Coef	SE Coef	T	P
Constant	31.334	4.180	7.50	0.000
AgeCo	0.22070	0.06099	3.62	0.000
AgeCo**2	-0.0001437	0.0002121	-0.68	0.499

S = 3.604   R-Sq = 77.7%   R-Sq(adj) = 77.4%

Regression Analysis: HeightCo versus AgeCo

The regression equation is
HeightCo = 34.1 + 0.180 AgeCo

158 cases used 28 cases contain missing values

Predictor	Coef	SE Coef	T	P
Constant	34.060	1.136	29.97	0.000
AgeCo	0.179700	0.007717	23.29	0.000

S = 3.597   R-Sq = 77.7%   R-Sq(adj) = 77.5%

Regression Analysis: WeightCo versus AgeCo, AgeCo2**

The regression equation is
WeightCo = 23.8 + 0.180 AgeCo + 0.00229 AgeCo**2

165 cases used 21 cases contain missing values

Predictor	Coef	SE Coef	T	P
Constant	23.78	23.55	1.01	0.314
AgeCo	0.1799	0.3435	0.52	0.601
AgeCo**2	0.002292	0.001195	1.92	0.057

S = 20.84   R-Sq = 69.3%   R-Sq(adj) = 68.9%

Regression Analysis: WeightCo versus AgeCo

The regression equation is
WeightCo = - 19.7 + 0.833 AgeCo

165 cases used 21 cases contain missing values

Predictor	Coef	SE Coef	T	P
Constant	-19.659	6.507	-3.02	0.003
AgeCo	0.83340	0.04414	18.88	0.000

S = 21.01   R-Sq = 68.6%   R-Sq(adj) = 68.4%

Regression Analysis: ln(WeightCo) versus AgeCo

The regression equation is
ln(WeightCo) = 3.30 + 0.00864 AgeCo

165 cases used 21 cases contain missing values

Predictor	Coef	SE Coef	T	P
Constant	3.29653	0.05685	57.99	0.000
AgeCo	0.0086442	0.0003857	22.41	0.000

S = 0.1836   R-Sq = 75.5%   R-Sq(adj) = 75.4%

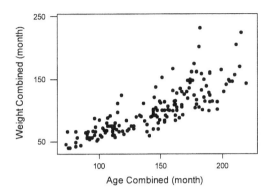

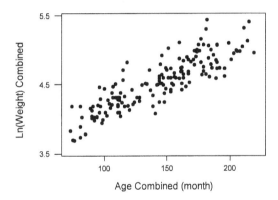

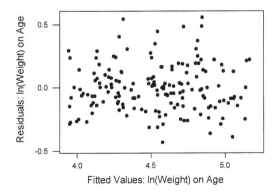

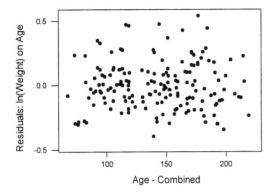

The Box-Cox transformation is applied to the response (see Section 6.5 in Chapter 6). For various values of λ we calculate the geometric mean $\bar{y}_g = (\Pi y_i)^{1/n}$ and the transformed response $(y^\lambda - 1)/\lambda(\bar{y}_g)^{\lambda-1}$, regress the transformed response on the explanatory variable AgeCo, and compute the error sum of squares $SSE(\lambda)$. The maximum likelihood estimate of λ minimizes $SSE(\lambda)$. The graph of $SSE(\lambda)$ against λ (given below) shows that the estimate of λ is close to 0. This confirms that the logarithmic transformation is appropriate.

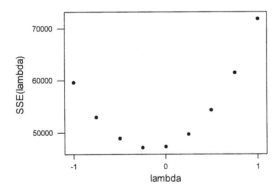

Part 2(a):
A plot of the weight against the height of mothers shows a relationship (correlation coefficient r = 0.336). A correlation coefficient of 0.34 implies that (only) about ten percent of the variability in weight is explained by height (because in simple linear

46

regression, $R^2 = r^2$). A similar conclusion can be reached for fathers. A plot of the weight against the height of fathers shows a similar-sized correlation (correlation coefficient $r = 0.289$).

Part 2(b):
The correlation between the height of mothers and the height of fathers is small ($r = 0.077$).
The correlation between the weight of mothers and the weight of fathers is larger (0.242). There is some (but rather weak) evidence that both partners tend to be above or below the average weight. The scatter plot shows three unusual cases. In one case the father is quite heavy, while the mother is of average weight. In the other two cases the fathers are of average weight while mothers have weights much above average. However, the omission of these three cases does not change the correlation coefficient ($r = 0.243$).

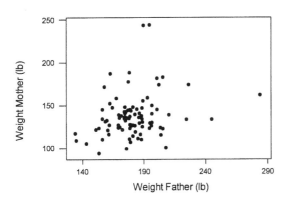

CHAPTER 9

A note on computing with SPSS (Version 11.5):

We use the SPSS software to fit the nonlinear regression models of Chapter 9. SPSS works like a spreadsheet program. We enter the data into the various columns of the spreadsheet and use the tabs: Analyze > Regression > Nonlinear. We write out the model equation and specify initial parameter values. We can save the fitted values and the residuals (also the derivatives of the objective function) into columns of the worksheet.

Several options for the iterative nonlinear estimation procedure are available. In the following examples we have used the Levenberg-Marquardt algorithm. Options for specifying the number of iterations and various convergence cutoffs are available. See the SPSS on-line help for further discussion and examples.

9.1 A graph of the leaf area against the age of the palm tree is given below.

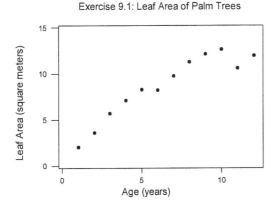

Note that there is not an abundance of data points to determine the model. The graph indicates that the relationship between leaf area and age is not linear; a quadratic component needs to be added to the model. The estimation results for the quadratic model $y = \beta_0 + \beta_1 \text{Age} + \beta_2 \text{Age}^2 + \varepsilon$ (Minitab output) is shown below. The quadratic coefficient is clearly needed; the estimate of the coefficient for Age**2 is -0.09616, with a significant t-ratio of -4.95.

Regression Analysis: Area (square meters) versus Age, Age**2

```
The regression equation is
Area (square meters) = - 0.123 + 2.15 Age - 0.0962 Age**2
```

Predictor	Coef	SE Coef	T	P
Constant	-0.1234	0.7334	-0.17	0.870
Age	2.1496	0.2594	8.29	0.000
Age**2	-0.09616	0.01942	-4.95	0.001

```
S = 0.7096     R-Sq = 96.6%     R-Sq(adj) = 95.8%
```

Analysis of Variance

Source	DF	SS	MS	F	P
Regression	2	128.071	64.036	127.19	0.000
Residual Error	9	4.531	0.503		
Total	11	132.603			

Rasch/Sedlacek use the Gompertz model $y = \mu + \varepsilon = \alpha \exp[-\beta \exp(-\gamma \text{Age})] + \varepsilon$ with parameters $\alpha > 0, \beta > 0, \gamma > 0$. Before fitting this model, we need to determine suitable starting values for the iterative nonlinear parameter estimation. The graph indicates that the saturation level for large values of Age is about 15. Hence a suitable starting value for α is given by 15. For Age = 1, the response is about 2; for Age = 5, the response is roughly 7. The model equation implies $-\beta \exp(-\gamma) = \ln(2/15)$ and $-\beta \exp(-5\gamma) = \ln(7/15)$. This implies $\exp(4\gamma) = [\ln(2/15)]/[\ln(7/15)]$ and $\gamma = \{\ln[\ln(2/15))/\ln(7/15)]\}/4 \approx 0.25$. Finally, $-\beta \exp(-\gamma) = \ln(2/15)$ and $\beta = -\ln(2/15)\exp(\gamma) \approx 2.6$. The starting values $\alpha = 15$, $\beta = 2.6$ and $\gamma = 0.25$ are used in the SPSS nonlinear regression routine. The (SPSS) outcome is given below:

Iteration	Residual SS	A	B	C
1	12.59000092	15.0000000	2.60000000	.250000000
1.1	15.64377972	11.3812687	2.34691685	.336739045
1.2	6.515778841	13.4641276	2.14122037	.271436482
2	6.515778841	13.4641276	2.14122037	.271436482
2.1	6.243186484	12.0109653	2.42204992	.359733910
3	6.243186484	12.0109653	2.42204992	.359733910
3.1	5.136619171	12.4921144	2.50012161	.359000316
4	5.136619171	12.4921144	2.50012161	.359000316
4.1	5.136518308	12.4937910	2.49764737	.358922047
5	5.136518308	12.4937910	2.49764737	.358922047
5.1	5.136518286	12.4936881	2.49773050	.358935226

```
Run stopped after 11 model evaluations and 5 derivative evaluations.
Iterations have been stopped because the relative reduction between
successive residual sums of squares is at most SSCON = 1.000E-08
```

```
Nonlinear Regression Summary Statistics      Dependent Variable AREA

  Source                   DF  Sum of Squares  Mean Square

  Regression                3    1023.43418     341.14473
  Residual                  9       5.13652         .57072
  Uncorrected Total        12    1028.57070
  (Corrected Total)        11     132.60269

  R squared = 1 - Residual SS / Corrected SS =     .96126

                                           Asymptotic 95 %
                              Asymptotic  Confidence Interval
  Parameter   Estimate       Std. Error    Lower        Upper

  A ($\alpha$)   12.493688057   .683789772 10.946848127 14.040527986
  B ($\beta$)     2.497730497   .440644079  1.500924338  3.494536656
  C ($\gamma$)     .358935226   .066769083   .207893067   .509977385

Asymptotic Correlation Matrix of the Parameter Estimates

                  A          B          C

  A            1.0000     -.4983     -.8306
  B            -.4983     1.0000      .8339
  C            -.8306      .8339     1.0000
```

The estimate of α is 12.5; the estimate of β is 2.5, and the estimate of γ is 0.36. All estimates are statistically significant. There is a fair amount of correlation, especially between the estimates of γ and α (-0.83) and the estimates of γ and β (0.83). The coefficient of determination (0.961) is similar to the R^2 from the quadratic regression. There is little difference between the fits of the quadratic regression (which is linear in the parameters) and the Gompertz model (which is nonlinear in the parameters). Both models lead to similar fitted curves. One difference is that the fitted values for the Gompertz model increase with age to an asymptotic value, whereas the quadratic curve starts to decrease with age after having reached a maximum. However, over the observed age range the two fitted models are virtually indistinguishable.

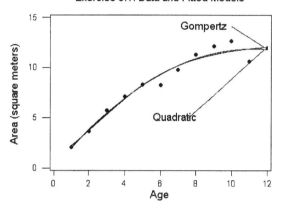

Exercise 9.1: Data and Fitted Models

9.2 A scatter plot of nitrate utilization versus light intensity is shown below. We use solid circles for day 1 observations, and triangles for day 2 observations. Furthermore, we have added some jitter to the light intensity in order to emphasize the differences between the measurements of day 1 and day 2. The day 2 measurements are slightly lower, especially at increasing light intensity.

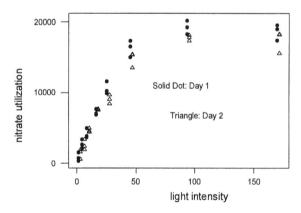

Exercise 9.2: Plot of nitrate utilization against light intensity

<u>Michaelis-Menton model:</u> Nitrate utilization reaches an asymptote of about 20,000 for large light intensity. Letting x go to infinity in the model equation

$$\frac{\beta_1 x}{\beta_2 + x} = \frac{\beta_1}{1 + (\beta_2 / x)} \approx 20{,}000$$

leads to the starting value $\beta_1 \approx 20,000$. Furthermore, the average nitrate utilization at light intensity 2.2 is 1075. Solving the model equation with $\beta_1 = 20,000$ leads to the starting value $\beta_2 = 38.7$.

Using these starting values in the SPSS nonlinear regression routine results in the following estimation results:

```
Nonlinear Regression Summary Statistics      Dependent Variable NITRATE

   Source                 DF  Sum of Squares  Mean Square

   Regression              2   6467226758.31  3233613379.15
   Residual               46   96536195.6932  2098612.94985
   Uncorrected Total      48   6563762954.00
   (Corrected Total)      47   2076766799.92

   R squared = 1 - Residual SS / Corrected SS =      .95352

                                           Asymptotic 95 %
                              Asymptotic    Confidence Interval
   Parameter    Estimate     Std. Error    Lower         Upper

   B1       23582.527043 889.35646658 21792.345325 25372.708760
   B2          34.243774004   3.427314571 27.344947587 41.142600421

   Asymptotic Correlation Matrix of the Parameter Estimates

                   B1         B2

   B1          1.0000      .8785
   B2           .8785     1.0000
```

Exponential rise model: Nitrate utilization reaches an asymptote of about 20,000 for large light intensity. Letting x go to infinity in the equation for the exponential rise model leads to the starting value $\beta_1 \approx 20,000$. The average nitrate utilization at light intensity 2.2 is 1075. Solving the model equation with $\beta_1 = 20,000$ leads to the starting value $\beta_2 = -\dfrac{1}{2.2}\ln\left[1-\dfrac{1075}{20,000}\right] = 0.025$. Using these starting values in the SPSS nonlinear regression program results in the estimation results:

```
Nonlinear Regression Summary Statistics      Dependent Variable NITRATE

   Source                 DF  Sum of Squares  Mean Square

   Regression              2   6504309173.87  3252154586.93
   Residual               46   59453780.1310  1292473.48111
   Uncorrected Total      48   6563762954.00
   (Corrected Total)      47   2076766799.92
```

```
R squared = 1 - Residual SS / Corrected SS =      .97137

                                       Asymptotic 95 %
                         Asymptotic   Confidence Interval
Parameter    Estimate    Std. Error   Lower        Upper

B1        19014.305975 398.04663684 18213.079652 19815.532299
B2           .030021624   .001629334   .026741945   .033301303

Asymptotic Correlation Matrix of the Parameter Estimates

                 B1          B2

B1           1.0000      -.7393
B2           -.7393      1.0000
```

Quadratic Michaelis-Menton model: Starting with $\beta_1 = 20,000$ and $\beta_2 = 38.7$ (from the earlier Michaelis-Menton model) and a small value for the parameter in the quadratic component ($\beta_3 = 0.1$) leads to the following results:

```
Nonlinear Regression Summary Statistics      Dependent Variable NITRATE

Source                 DF   Sum of Squares   Mean Square

Regression              3   6520540397.33   2173513465.78
Residual               45   43222556.6654    960501.25923
Uncorrected Total      48   6563762954.00
(Corrected Total)      47   2076766799.92

R squared = 1 - Residual SS / Corrected SS =      .97919

                                       Asymptotic 95 %
                         Asymptotic   Confidence Interval
Parameter    Estimate    Std. Error   Lower        Upper

B1        66769.034924 17585.504714 31350.010284 102188.05956
B2          137.82679758 43.735712594 49.738550634 225.91504453
B3            .011281055   .004496402   .002224837   .020337274


Asymptotic Correlation Matrix of the Parameter Estimates

                 B1          B2          B3

B1           1.0000      .9964       .9941
B2            .9964     1.0000       .9856
B3            .9941      .9856      1.0000
```

Modified exponential rise model: Using $\beta_1 = 20,000$ and $\beta_2 = 0.025$ from the earlier exponential rise model and a small value for $\beta_3 = 0.01$ leads to the following results:

```
Nonlinear Regression Summary Statistics      Dependent Variable NITRATE

   Source                   DF  Sum of Squares  Mean Square

   Regression                3  6519117089.28  2173039029.76
   Residual                 45  44645864.7154   992130.32701
   Uncorrected Total        48  6563762954.00
   (Corrected Total)        47  2076766799.92

   R squared = 1 - Residual SS / Corrected SS =    .97850

                                         Asymptotic 95 %
                           Asymptotic   Confidence Interval
   Parameter   Estimate    Std. Error    Lower        Upper

   B1      33551.454219 9502.1687711 14413.103896 52689.804543
   B2          .018534079   .003572151   .011339397   .025728761
   B3          .003221159   .001338559   .000525162   .005917155

   Asymptotic Correlation Matrix of the Parameter Estimates

                   B1          B2          B3

   B1          1.0000      -.9898       .9948
   B2          -.9898      1.0000       -.9741
   B3           .9948      -.9741      1.0000
```

All four models lead to large R^2. The Michaelis-Menton and its quadratic extension lead to R^2 of 0.954 and 0.979, respectively. Carrying out an F-test for the significance of the quadratic component in the Michaelis-Menton model leads to the F-statistic $F = [96,536,195 - 43,222,556]/[43,222,556/45] = 55.5$, which is highly significant. This shows that the quadratic extension represents a significant improvement.

Similarly, the exponential rise model and its extension lead to R^2 of 0.971 and 0.979, respectively. The F-test for the significance of the extra component in the exponential rise model leads to the F-statistic $F = [59,453,780 - 44,645,864]/[44,645,864/45] = 14.9$, which is also highly significant.

The extensions are beneficial. The modified Michaelis-Menton and the modified exponential rise models perform similarly. In the following graph we show the fit of the quadratic Michaelis-Menton model; the fitted values of the modified exponential rise model are virtually indistinguishable.

Exercise 9.2: Fit of the quadratic Michaelis-Menten model

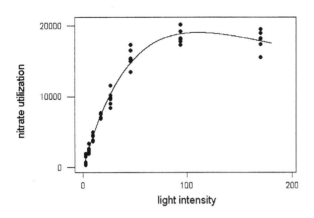

Standard Michaelis-Menton model with an indicator for the change of day: The final parameter estimates in the previous Michaelis-Menton model, $\hat{\beta}_1 = 23,500$ and $\hat{\beta}_2 = 34.2$, are taken as the starting values in the iterative nonlinear estimation. Small values for the day indicator $\alpha_1 = -1000, \alpha_2 = -1$ are used as the starting values for the two additional parameters. The estimation results are given below:

```
Nonlinear Regression Summary Statistics    Dependent Variable NITRATE

  Source                    DF  Sum of Squares  Mean Square

  Regression                 4  6477253424.57  1619313356.14
  Residual                  44  86509529.4274  1966125.66881
  Uncorrected Total         48  6563762954.00
  (Corrected Total)         47  2076766799.92

  R squared = 1 - Residual SS / Corrected SS =      .95834

                                         Asymptotic 95 %
                         Asymptotic   Confidence Interval
  Parameter   Estimate   Std. Error    Lower        Upper

  B1      24743.334444 1241.1211323 22242.019158 27244.649730
  B2         35.275400267  4.656586052 25.890667730 44.660132803
  A1      -2328.743446 1720.3472191 -5795.875448 1138.3885567
  A2         -2.172827290  6.626226364 -15.52710905 11.181454466
```

```
Asymptotic Correlation Matrix of the Parameter Estimates

                  B1          B2          A1          A2

    B1        1.0000       .8810      -.7214      -.6191
    B2         .8810      1.0000      -.6356      -.7028
    A1        -.7214      -.6356      1.0000       .8781
    A2        -.6191      -.7028       .8781      1.0000
```

The F-statistic for testing the null hypothesis $\alpha_1 = \alpha_2 = 0$ is
$F = [(96{,}536{,}195 - 86{,}509{,}529)/2]/[86{,}509{,}529/44] = 2.55$. The probability value
from the F(2,44) distribution is $P[F(2,44) \geq 2.55] = 1 - 0.91 = 0.09$. Hence there is
only weak evidence for including a day effect. The individual confidence intervals for
α_1 and α_2 cover zero, which makes the individual interpretation of the two day-effect
parameters difficult. These estimates are also quite correlated.

Quadratic Michaelis-Menton model with an indicator for the change of day: The final
values from the earlier quadratic model $\hat{\beta}_1 = 66{,}700, \hat{\beta}_2 = 138, \hat{\beta}_3 = 0.01$ and small
values for the three parameters associated with the day indicators,
$\alpha_1 = -2000, \alpha_2 = -2, \alpha_3 = 0.001$, are used as the starting values in the iterative
nonlinear SPSS estimation. The estimation results are given below:

```
Run stopped after 10 model evaluations and 5 derivative evaluations.
Iterations have been stopped because the relative reduction between
successive residual sums of squares is at most SSCON = 1.000E-08


Nonlinear Regression Summary Statistics     Dependent Variable NITRATE

    Source                  DF  Sum of Squares  Mean Square

    Regression               6  6531740362.05  1088623393.67
    Residual                42  32022591.9535   762442.66556
    Uncorrected Total       48  6563762954.00
    (Corrected Total)       47  2076766799.92

  R squared = 1 - Residual SS / Corrected SS =      .98458

                                          Asymptotic 95 %
                            Asymptotic   Confidence Interval
    Parameter    Estimate   Std. Error    Lower         Upper

    B1        89797.916970 37540.345749 14038.432096 165557.40184
    B2        186.61862445 89.984553967  5.022442558 368.21480635
    A1        -38897.78690 39982.748874 -119586.2408 41790.667033
    A2        -83.09078151 96.727346453 -278.2944695 112.11290653
    B3          .016252421   .009207288  -.002328638   .034833481
    A3         -.008449660   .009916211  -.028461384   .011562065
```

57

Asymptotic Correlation Matrix of the Parameter Estimates

	B1	B2	A1	A2	B3	A3
B1	1.0000	.9978	-.9389	-.9283	.9965	-.9252
B2	.9978	1.0000	-.9369	-.9303	.9913	-.9204
A1	-.9389	-.9369	1.0000	.9971	-.9356	.9953
A2	-.9283	-.9303	.9971	1.0000	-.9222	.9894
B3	.9965	.9913	-.9356	-.9222	1.0000	-.9285
A3	-.9252	-.9204	.9953	.9894	-.9285	1.0000

The F-statistic for testing the null hypothesis $\alpha_1 = \alpha_2 = \alpha_3 = 0$ is given by $F = [(43{,}222{,}556 - 32{,}022{,}591)/3]/[32{,}022{,}591/42] = 4.90$. The probability value from the $F(3,42)$ distribution is $P[F(3,42) \geq 4.90] = 1 - 0.995 = 0.005$, showing that the indicators for the day effect help explain the variation. Individually the three parameters are statistically insignificant and also highly correlated. This makes an individual interpretation of the estimates difficult.

The graph shown below compares the quadratic Michaelis-Menton model with and without the day indicator. The graph shows that the quadratic Michaelis-Menton model with a day indicator is capable of expressing the day differences.

Exercise 9.2: Quadratic Michaelis-Menton model with day indicator

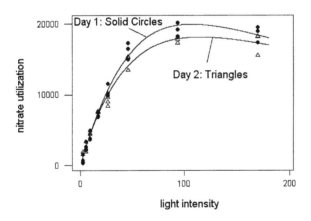

9.3

Model 1: The logarithmic transformation of the first model leads to
$$\ln(y) = \ln(\beta_0) + \beta_1 \ln(x_1) + \beta_2 \ln(x_2) + \ln(\varepsilon)$$
A standard multiple linear regression of $\ln(y)$ on $\ln(x_1)$ and $\ln(x_2)$ leads to the estimates of $\alpha = \ln(\beta_0), \beta_1$ and β_2. The estimate of β_0 can be obtained from

$\beta_0 = \exp(\alpha)$. When carrying out the regression with the transformed variables we need to assume that the error $\ln(\varepsilon)$ satisfies the standard regression assumptions.

<u>Model 2:</u> Taking the reciprocal of the response in the second model leads to

$$1/y = \beta_0 + \beta_1 x + \varepsilon$$

A simple linear regression of $(1/y)$ on x_1 leads to the estimates of β_0, β_1.

<u>Model 3:</u> The reciprocal of the response and a subsequent logarithmic transformation leads to the model

$$\ln[(1/y) - 1] = \beta_0 + \beta_1 x_1 + \ln(\varepsilon)$$

A simple linear regression of $\ln[(1/y) - 1]$ on x_1 leads to the estimates of β_0, β_1. We need to assume that the error $\ln(\varepsilon)$ satisfies the standard regression assumptions.

CHAPTER 10

A note on computing in time series situations

The **Minitab** software is used here for calculating the autocorrelation function of time series observations and for fitting the autoregressive integrated moving average (ARIMA) models in Chapter 10. The class of ARIMA models includes the autoregressive, random walk, and noisy random walk models discussed in Chapter 10. The Minitab ARIMA routine also facilitates the computation of the predictions and prediction intervals.

Combined regression time series models can be estimated within the **SCA** software or within the econometrics software **EVIEWS**. Contact information for these two software providers are:

- SCA: Scientific Computing Associates Corp.,1410 N. Harlem Avenue, River Forest, IL 60305. www.scausa.com.
- EVIEWS: QMS (Quantitative Micro Software), 4521 Campus Drive, Irvine, CA, 92612. www.eviews.com

For SCA one needs to construct a text file macro which is then executed by the software. The output can be saved into a file. Here we list the text file macro for Exercise 10.13.

```
==MACRO
Input variables are year quarter FTEShare Car FTEComm.
        1952  3     112.7 105761     96.21
        1952  4     115.0 121874     93.74
        1953  1     121.4 126260     91.37
        ...

        ...
        1967  2     343.1 393808     79.90
        1967  3     360.8 375968     78.70
        1967  4     397.8 381692     81.50
end
print variables are year quarter FTEShare Car FTEComm.
Utsmodel name is m1. @
Model is FTEShare((1-B)) = (w1*B**6)Car((1-B)) @
+ (w2*B**7)FTEComm((1-B)) + (1-theta*B)noise.
        Model m1 considers the differences of the response and the regressor
        variables. The regression model relates the differences of the response to the
        differences of Car (with lag 6) and the differences of FTECom (with lag 7). A
        first order moving average model is taken as the error model.
Uestim m1. Method is EXACT. Hold residuals(resid1).
```

```
Acf variable is resid1.
Utsmodel name is m2. @
Model is FTEShare((1-B)) = (w1*B**6)Car((1-B)) @
+ (w2*B**7)FTEComm((1-B)) + 1/(1-phi*B)noise.
```
> Model m2 considers differences of the response and the regressor variables. A
> first order autoregressive model is used as the error model.

```
Uestim m2. Method is EXACT. Hold residuals(resid2).
Acf variable is resid2.
RETURN
```

Many options are available within SCA. See the SCA on-line help for further
discussion and examples.

A short primer on the backshift operator

The <u>backshift operator</u> B simplifies the notation of time series models. When applied to a time series y_t, the backshift operator shifts the time index by one unit. That is,

$$By_t = y_{t-1}, B^2y_t = y_{t-2}, B^3y_t = y_{t-3}, \text{ and so on.}$$

Similarly,

$$Bx_t = x_{t-1}, B^2x_t = x_{t-2}, B^3x_t = x_{t-3}, \text{ and so on.}$$

<u>First differences</u> of a time series can be written as $y_t - y_{t-1} = y_t - By_t = (1-B)y_t$.

<u>Second differences</u> (the difference of differences) as

$$y_t - y_{t-1} - (y_{t-1} - y_{t-2}) = (1-B)y_t - (1-B)y_{t-1} = (1-B)(y_t - y_{t-1}) = (1-B)^2 y_t$$

The first order <u>moving average model</u> can be written as

$$\varepsilon_t = a_t - \theta a_{t-1} \quad \text{or} \quad \varepsilon_t = (1-\theta B)a_t.$$

The first order <u>autoregressive model</u> can be written as

$$\varepsilon_t = \phi\varepsilon_{t-1} + a_t \quad \text{or} \quad \varepsilon_t - \phi B\varepsilon_t = a_t \quad \text{or} \quad (1-\phi B)\varepsilon_t = a_t.$$

We can also write it as

$$\varepsilon_t = \frac{1}{1-\phi B}a_t = (1 + \phi B + \phi^2 B^2 + ...)a_t = a_t + \phi a_{t-1} + \phi^2 a_{t-2} + ... \quad .$$

The <u>noisy random walk</u> also known as the ARIMA(0,1,1) model,

$\varepsilon_t - \varepsilon_{t-1} = a_t - \theta a_{t-1}$, can be written as $(1-B)\varepsilon_t = (1-\theta B)a_t$. Or, as $\varepsilon_t = \frac{1-\theta B}{1-B}a_t$.

<u>Regression models with (first-order) moving average errors</u>

$$y_t = \beta_0 + \beta_1 x_t + \varepsilon_t \quad \text{with} \quad \varepsilon_t = (1-\theta B)a_t$$

can be combined as

$$y_t = \beta_0 + \beta_1 x_t + (1-\theta B)a_t.$$

<u>Regression models with (first-order) autoregressive errors</u>

$$y_t = \beta_0 + \beta_1 x_t + \varepsilon_t \quad \text{with} \quad (1-\phi B)\varepsilon_t = a_t$$

can be combined as

$$y_t = \beta_0 + \beta_1 x_t + \frac{1}{1-\phi B}a_t.$$

<u>Regression models with noisy random walk errors</u>

$$y_t = \beta_0 + \beta_1 x_t + \varepsilon_t \quad \text{with} \quad (1-B)\varepsilon_t = (1-\theta B)a_t$$

can be combined as

$$y_t = \beta_0 + \beta_1 x_t + \frac{1 - \theta B}{1 - B} a_t.$$

Alternatively, this model can be written as a regression of differences,

$$(1 - B)y_t = \beta_1 (1 - B)x_t + (1 - \theta B)a_t;$$

the constant disappears as $(1 - B)\beta_0 = \beta_0 - \beta_0 = 0$.

10.3 (a) The time series plot of the data is given below.

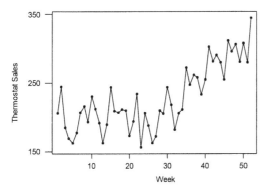

Exercise 10.3: Time series plot of weekly thermostat sales

(b) The MINITAB output of the regression of sales on time, $y_t = \beta_0 + \beta_1 t + \varepsilon_t$, is shown below. The predictions and the 95 percent prediction intervals for the next three observations are calculated from the results in Section 4.3.2.

```
The regression equation is
Sales = 166 + 2.32 Time

Predictor          Coef      SE Coef            T         P
Constant        166.396        8.760        19.00     0.000
Time             2.3247        0.2876         8.08     0.000

S = 31.13        R-Sq = 56.6%      R-Sq(adj) = 55.8%

Analysis of Variance

Source             DF           SS           MS           F         P
Regression          1        63299        63299       65.32     0.000
Residual Error     50        48451          969
Total              51       111750
```

Prediction for the next period (time = 53):
Prediction: $y_{52}(1) = 166.396 + (2.325)(53) = 289.60$

Prediction interval: 224.65, 354.56
Predictions and prediction intervals can be obtained with the Minitab option in the "regress" command. Alternatively, one can calculate them from the results in Chapter 2,

$$y_{52}(1) \pm 2.0086\sqrt{969}\sqrt{1 + \frac{1}{52} + \frac{(53-26.5)^2}{11,713}},$$

where 2.0086 is the 97.5th percentile of the t-distribution with 50 degrees of freedom, $26.5 = (1/52)\sum_{t=1}^{52} t$ and $11{,}713 = \sum_{t=1}^{52} (t-26.5)^2$.

Prediction for two periods ahead (time = 54):
Prediction: $y_{52}(2) = 166.396 + (2.325)(54) = 291.93$
Prediction interval: 226.84, 357.02

$$y_{52}(2) \pm 2.0086\sqrt{969}\sqrt{1 + \frac{1}{52} + \frac{(54-26.5)^2}{11{,}713}}$$

Prediction for three periods ahead (time = 55):
Prediction: $y_{52}(3) = 166.396 + (2.325)(55) = 294.25$
Prediction interval: 229.02, 359.49

$$y_{52}(3) \pm 2.0086\sqrt{969}\sqrt{1 + \frac{1}{52} + \frac{(55-26.5)^2}{11{,}713}}$$

(c) The Durbin-Watson test statistic is 1.09, and far from the desired value 2. It is not acceptable. There is autocorrelation in the residuals. The first ten autocorrelations are given below (read across):

```
0.405962   0.257128   0.184543   0.192049   0.274287
0.401941   0.283462   0.172746   0.091004  -0.070815
```

The approximate standard error of an autocorrelation is given by $1/\sqrt{52} = 0.14$. Several of the autocorrelations exceed twice the standard error. The autocorrelations tend to be positive with a slow decay, indicating an autocorrelation problem and possible nonstationarity. A regression of sales on time, $y_t = \beta_0 + \beta_1 t + \varepsilon_t$, is definitely not an appropriate forecasting model. The plot of the residuals against time (given below) shows patterns.

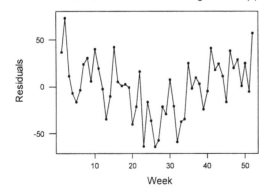

Exercise 10.3: Residuals from the regression in (a)

(d) The mean of the first differences is 2.7255. This becomes the estimate of β_1 in the model $\Delta y_t = \beta_1 + a_t$. The standard deviation of the first differences is 32.51; this becomes the estimate of σ_a.

The forecasts for the next three observations are:
$$y_{52}(1) = y_{52} + \hat{\beta}_1 = 345 + 2.73 = 347.73$$
$$y_{52}(2) = y_{52}(1) + \hat{\beta}_1 = 347.73 + 2.73 = 350.46$$
$$y_{52}(3) = y_{52}(2) + \hat{\beta}_1 = 350.46 + 2.73 = 353.19$$

The prediction intervals are given by

$y_{52}(1) \pm (1.96)(32.51)$	or	347.73 ± 63.72
$y_{52}(2) \pm (1.96)(\sqrt{2})(32.51)$	or	350.46 ± 90.11
$y_{52}(3) \pm (1.96)(\sqrt{3})(32.51)$	or	353.19 ± 110.37

The first ten autocorrelations of the differenced series are given below (read across):

```
-0.365082   -0.059187   -0.033625   -0.093252   -0.041308
 0.186040    0.048240   -0.038622    0.034502   -0.169835
```

The lag one autocorrelation exceeds twice its approximate standard error $1/\sqrt{51} = 0.14$. Hence this is <u>not</u> an appropriate forecasting model.

(e) The ARIMA time series procedure in MINITAB is used to estimate the noisy random walk model $\Delta y_t = y_t - y_{t-1} = \beta_1 + a_t - \theta a_{t-1}$. Using the MINITAB ARIMA command, we find

```
Estimates at each iteration
Iteration      SSE      Parameters
    0        49361.5    0.100    2.825
    1        45310.4    0.250    2.496
    2        42249.3    0.400    2.245
    3        39884.7    0.550    2.106
    4        38533.0    0.687    2.124
    5        38448.9    0.717    2.220
    6        38447.7    0.719    2.248
    7        38447.7    0.720    2.251
    8        38447.7    0.720    2.252
Relative change in each estimate less than  0.0010

Final Estimates of Parameters
Type         Coef     SE Coef        T        P
MA    1      0.7198    0.1010      7.13    0.000
Constant    2.252      1.127       2.00    0.051
```

```
Differencing: 1 regular difference
Number of observations:  Original series 52, after differencing 51
Residuals:     SS =   38356.2  (backforecasts excluded)
               MS =    782.8  DF = 49


Forecasts from period 52
                              95 Percent Limits
Period       Forecast        Lower          Upper
  53         313.544        258.696        368.392
  54         315.796        258.836        372.756
  55         318.048        259.052        377.045
```

The estimates are $\hat{\beta}_1 = 2.252$ and $\hat{\theta} = 0.72$. The forecasts and the 95 percent prediction intervals are part of the MINITAB output. The first ten autocorrelations of the residuals from this model are shown below. They are small (most of them smaller than their standard error), indicating that we have found an acceptable model.

```
0.066442   -0.067055   -0.127384   -0.104795    0.045999
0.283976    0.172438    0.061706   -0.010849   -0.161526
```

10.4 (a) The time series plot shows that the linear trend is not globally stable. The trend shifts over time. Hence a regression on time, $y_t = \beta_0 + \beta_1 t + \varepsilon_t$, is not appropriate. The residuals from the (incorrect) regression on time show (positive) autocorrelations and an unacceptable Durbin-Watson test statistic (0.26) that is considerably smaller than 2.

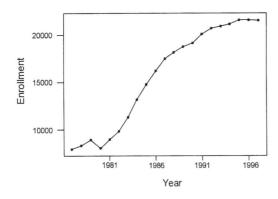

Exercise 10.4: Enrollment

```
The regression equation is
enrollment = 6527 + 830 time

Predictor        Coef        SE Coef           T          P
Constant       6527.2          599.6       10.89      0.000
time          830.08           47.75       17.38      0.000
```

68

```
S = 1325        R-Sq = 94.1%     R-Sq(adj) = 93.8%
```

Analysis of Variance

```
Source            DF        SS          MS          F        P
Regression        1     530560905   530560905    302.14    0.000
Residual Error    19     33363694    1755984
Total             20    563924600
```

Durbin-Watson statistic = 0.26

First four autocorrelations of residuals

```
   0.779040    0.504676    0.191752   -0.088873
```

The predictions and 95% prediction intervals for the next three periods are given below. Because of the residual problems with this model, these predictions should not be used:

For the next period (time = 22): 24,789 and (21,745 to 27,833)
For two periods ahead (time = 23): 25,619 and (22,537 to 28,701)
For three periods ahead (time = 24): 26,449 and (23,327 to 29,571)

(b) The mean of the first differences is 682. This becomes the estimate of β_1. The standard deviation of the first differences is 654; this becomes the estimate of σ_ε.

The forecasts for the next three observations are:

$$y_{21}(1) = y_{21} + \hat{\beta}_1 = 21,531 + 682 = 22,213$$
$$y_{21}(2) = y_{21}(1) + \hat{\beta}_1 = y_{21} + 2\hat{\beta}_1 = 22,213 + 682 = 22,895$$
$$y_{21}(3) = y_{21}(2) + \hat{\beta}_1 = y_{21} + 3\hat{\beta}_1 = 22,895 + 682 = 23,577$$

The prediction intervals are given by

$y_{21}(1) \pm (1.96)(654)$	or	$22,213 \pm 1,282$
$y_{21}(2) \pm (1.96)(\sqrt{2})(654)$	or	$22,895 \pm 1,813$
$y_{21}(3) \pm (1.96)(\sqrt{3})(654)$	or	$23,577 \pm 2,220$

The first four autocorrelations of the differenced series are given below (read across):

```
    0.491156    0.393677    0.114746   -0.074641
```

The lag one autocorrelation exceeds twice its approximate standard error $1/\sqrt{20} = 0.22$.

This forecasting model is not appropriate.

(c) The regression of enrollment on the previous two enrollments (lag one and two), $y_t = \beta_0 + \phi_1 y_{t-1} + \phi_2 y_{t-2} + \varepsilon_t$, is given below. The Durbin-Watson statistic is much better; it is close to the desired value 2. Also, the autocorrelations of the residuals are small. This model provides an appropriate forecasting method.

```
The regression equation is
enroll = 914 + 1.47 enroll-1 - 0.506 enroll-2

19 cases used 2 cases contain missing values

Predictor        Coef     SE Coef          T        P
Constant        914.4       477.6       1.91    0.074
enroll-1       1.4691      0.2147       6.84    0.000
enroll-2      -0.5061      0.2108      -2.40    0.029

S = 575.2       R-Sq = 98.8%     R-Sq(adj) = 98.6%

Analysis of Variance

Source           DF          SS          MS        F        P
Regression        2   431791259   215895629   652.54    0.000
Residual Error   16     5293676      330855
Total            18   437084935

Durbin-Watson statistic = 2.32

First four autocorrelations of the residuals:

  -0.168526    0.104467   -0.054733   -0.121096
```

The root mean square error from the second-order autoregression, $\sqrt{330{,}855} = 575$, is considerably smaller than the root mean square error of the regression on time in (a), $\sqrt{1{,}755{,}984} = 1{,}325$. The AR(2) model is preferable.

The forecasts can be obtained from:

$$y_{21}(1) = 914 + 1.47 y_{21} - 0.51 y_{20} = 914 + 1.47(21{,}531) - 0.51(21{,}624) = 21{,}536$$
$$y_{21}(2) = 914 + 1.47 y_{21}(1) - 0.51 y_{21} = 914 + 1.47(21{,}536) - 0.51(21{,}531) = 21{,}592$$
$$y_{21}(3) = 914 + 1.47 y_{21}(2) - 0.51 y_{21}(1) = 914 + 1.47(21{,}592) - 0.51(21{,}536) = 21{,}670$$

Another reasonable model for these data is the second difference model,
$$(y_t - y_{t-1}) - (y_{t-1} - y_{t-2}) = y_t - 2y_{t-1} + y_{t-2} = \varepsilon_t \ .$$
It is a special case of the AR(2) model with $\phi_1 = 2$ and $\phi_2 = -1$. The forecasts are

$$y_{21}(1) = 2y_{21} - y_{20} = 2(21{,}531) - 21{,}624 = 21{,}438$$

$$y_{22}(2) = 2y_{21}(1) - y_{21} = 2(21{,}438) - 21{,}531 = 21{,}345$$
$$y_{21}(3) = 2y_{21}(2) - y_{21}(1) = 2(21{,}345) - 21{,}438 = 21{,}252$$

10.5 (a) A time series graph of the observations shows the high sales activity during December months. The question whether or not the data exhibit a trend component is difficult to answer from just the graph alone.

Exercise 10.5: Sales - Center City Bookstore

We consider a model with a linear time trend and monthly indicators that account for the seasonal pattern,
$$\text{Sales}_t = \beta_0 + \beta_1 t + \beta_2 \text{IndJan}_t + \beta_3 \text{IndFeb}_t + \ldots + \beta_{12}\text{IndNov}_t + \varepsilon_t \ .$$
The estimation results indicate a positive trend component. The probability value of the trend coefficient is 0.058, which indicates weak statistical significance. The magnitude of the trend coefficient, a 0.45 EURO increase per month, is of no practical importance. The coefficients of the indicators express differences in average sales for the various months and their base of comparison (December). For example, the value for January (-1,154) indicates that sales in January are on average 1,154 EUROs lower than those in December. The residuals from the regression are still autocorrelated, especially at lag 1; the lag one autocorrelation -0.23 exceeds twice its standard error, $1/\sqrt{94} = 0.10$. The Durbin-Watson statistic (2.45) is larger than 2, reflecting a negative lag one autocorrelation.

```
The regression equation is
Sales = 1500 + 0.449 Time - 1154 IndJan - 1169 IndFeb - 1073 IndMar
             - 1049 IndApr - 1057 IndMay - 1061 IndJun - 1126 IndJul
             - 1062 IndAug - 984 IndSep - 951 IndOct - 776 IndNov

Predictor         Coef     SE Coef         T        P
Constant       1500.18       25.68     58.42    0.000
Time            0.4487      0.2335      1.92    0.058
```

```
IndJan      -1154.47       31.66      -36.47      0.000
IndFeb      -1169.04       31.65      -36.94      0.000
IndMar      -1073.12       31.64      -33.91      0.000
IndApr      -1048.82       31.64      -33.15      0.000
IndMay      -1057.27       31.64      -33.42      0.000
IndJun      -1060.96       31.64      -33.54      0.000
IndJul      -1125.91       31.64      -35.59      0.000
IndAug      -1061.74       31.64      -33.56      0.000
IndSep       -983.94       31.64      -31.09      0.000
IndOct       -951.13       31.65      -30.05      0.000
IndNov       -776.41       32.67      -23.76      0.000

S = 61.13       R-Sq = 96.4%      R-Sq(adj) = 95.8%

Analysis of Variance

Source            DF          SS          MS          F        P
Regression        12      7992176      666015      178.25    0.000
Residual Error    81       302649        3736
Total             93      8294825

Durbin-Watson statistic = 2.45
```

Autocorrelation Function of Residuals

```
Lag Corr   T   LBQ    Lag Corr   T   LBQ    Lag Corr   T   LBQ    Lag Corr   T   LBQ
1 -0.23 -2.26  5.27    8 -0.13 -1.16 12.15   15 -0.14 -1.22 18.16   22 0.13 1.03 21.81
2 -0.05 -0.47  5.53    9  0.05  0.42 12.39   16  0.06  0.53 18.63   23 -0.00 -0.02 21.81
3  0.09  0.83  6.34   10 -0.03 -0.23 12.47   17  0.01  0.09 18.64
4 -0.16 -1.45  8.85   11  0.10  0.88 13.59   18 -0.00 -0.04 18.65
5  0.05  0.46  9.12   12 -0.10 -0.84 14.64   19  0.03  0.25 18.76
6 -0.11 -0.96 10.31   13 -0.03 -0.25 14.74   20  0.03  0.26 18.88
7 -0.02 -0.13 10.34   14  0.10  0.83 15.81   21 -0.09 -0.74 19.85
```

(b) The autocorrelation function of the residuals from the model in (a) has a spike at lag one. This suggests a first-order moving average model for the errors. Alternatively, one could consider a first-order autoregressive model. We study both error models and show that the results for these two error models are very similar.

$$\text{MA(1): Sales}_t = \beta_0 + \beta_1 t + \beta_2 \text{IndJan}_t + \beta_3 \text{IndFeb}_t + \ldots + \beta_{12} \text{IndNov}_t + (1 - \theta B) a_t$$

or,

$$\text{AR(1): Sales}_t = \beta_0 + \beta_1 t + \beta_2 \text{IndJan}_t + \beta_3 \text{IndFeb}_t + \ldots + \beta_{12} \text{IndNov}_t + \frac{1}{1 - \phi B} a_t$$

We use SCA to estimate the models (alternatively, one could use Eviews). The results for MA(1) errors are shown first. The residuals from the revised model are uncorrelated. The lag one autocorrelation of the residuals is 0.10, and is well within one standard error. The trend coefficient is small and can be neglected for practical purposes. The seasonal component is very strong.

PARAMETER LABEL	VARIABLE NAME	NUM./ DENOM.	FACTOR	ORDER	CONS-TRAINT	VALUE	STD ERROR	T VALUE
1 CNST		CNST	1	0	NONE	1500.8551	22.7676	65.92
2 B1	TIME	NUM.	1	0	NONE	.4363	.1545	2.82
3 B2	INDJAN	NUM.	1	0	NONE	-1154.6164	32.6848	-35.33
4 B3	INDFEB	NUM.	1	0	NONE	-1169.1776	29.3843	-39.79
5 B4	INDMAR	NUM.	1	0	NONE	-1073.2389	29.3796	-36.53
6 B5	INDAPR	NUM.	1	0	NONE	-1048.9253	29.3758	-35.71
7 B6	INDMAY	NUM.	1	0	NONE	-1057.3616	29.3727	-36.00
8 B7	INDJUN	NUM.	1	0	NONE	-1061.0479	29.3705	-36.13
9 B8	INDJUL	NUM.	1	0	NONE	-1125.9842	29.3691	-38.34
10 B9	INDAUG	NUM.	1	0	NONE	-1061.7955	29.3685	-36.15
11 B10	INDSEP	NUM.	1	0	NONE	-983.9818	29.3687	-33.50
12 B11	INDOCT	NUM.	1	0	NONE	-951.1681	29.3698	-32.39
13 B12	INDNOV	NUM.	1	0	NONE	-778.9498	33.9353	-22.95
14 THETA	SALES	MA	1	1	NONE	.2721	.0995	2.74

```
EFFECTIVE NUMBER OF OBSERVATIONS . .       94
R-SQUARE . . . . . . . . . . . . . .      0.966
RESIDUAL STANDARD ERROR. . . . . .   0.548881E+02

AUTOCORRELATIONS OF RESIDUALS

 1- 12    .01 -.04  .04 -.15 -.01 -.13 -.09 -.15  .01 -.00  .08 -.09
 ST.E.    .10  .10  .10  .10  .11  .11  .11  .11  .11  .11  .11  .11

13- 24   -.04  .06 -.12  .04  .03  .01  .04  .03 -.05  .11  .02 -.02
 ST.E.    .11  .11  .11  .11  .11  .11  .11  .11  .11  .11  .12  .12
```

The results for AR(1) errors (shown below) are similar:

PARAMETER LABEL	VARIABLE NAME	NUM./ DENOM.	FACTOR	ORDER	CONS-TRAINT	VALUE	STD ERROR	T VALUE
1 CNST		CNST	1	0	NONE	1501.8372	23.1090	64.99
2 B1	TIME	NUM.	1	0	NONE	.4248	.1740	2.44
3 B2	INDJAN	NUM.	1	0	NONE	-1151.3433	33.8071	-34.06
4 B3	INDFEB	NUM.	1	0	NONE	-1170.5413	28.6817	-40.81
5 B4	INDMAR	NUM.	1	0	NONE	-1073.4953	29.6779	-36.17
6 B5	INDAPR	NUM.	1	0	NONE	-1049.4245	29.4279	-35.66
7 B6	INDMAY	NUM.	1	0	NONE	-1057.7898	29.4837	-35.88
8 B7	INDJUN	NUM.	1	0	NONE	-1061.4784	29.4667	-36.02
9 B8	INDJUL	NUM.	1	0	NONE	-1126.4000	29.4818	-38.21
10 B9	INDAUG	NUM.	1	0	NONE	-1062.2005	29.4317	-36.09
11 B10	INDSEP	NUM.	1	0	NONE	-984.3751	29.6497	-33.20
12 B11	INDOCT	NUM.	1	0	NONE	-951.5499	28.7184	-33.13
13 B12	INDNOV	NUM.	1	0	NONE	-779.1028	33.7655	-23.07
14 PHI	SALES	D-AR	1	1	NONE	-.2369	.1014	-2.34

73

```
EFFECTIVE NUMBER OF OBSERVATIONS . .          93
R-SQUARE . . . . . . . . . . . . .          0.965
RESIDUAL STANDARD ERROR. . . . . .   0.553693E+02

AUTOCORRELATIONS OF RESIDUALS

  1- 12    -.02 -.09  .05 -.14  .01 -.10 -.08 -.14  .01  .01  .07 -.10
  ST.E.     .10  .10  .10  .10  .11  .11  .11  .11  .11  .11  .11  .11

 13- 24    -.04  .06 -.11  .04  .03  .01  .04  .02 -.05  .11  .02 -.03
  ST.E.     .11  .11  .11  .11  .11  .11  .11  .11  .11  .11  .12  .12
```

(c) A scatter plot of sales against advertising is shown below. Adding advertising expenditures to our earlier specification, we consider the model

$$\text{Sales}_t = \beta_0 + \beta_1 t + \beta_2 \text{IndJan}_t + \beta_3 \text{IndFeb}_t + \ldots + \beta_{12}\text{IndNov}_t + \beta_{13}\text{Adv}_t + (1-\theta B)a_t$$

The estimation results are given below. We find little evidence that advertising provides additional information. This finding can be explained by the fact that advertising is (partially) confounded with the seasonal pattern represented by the seasonal indicators.

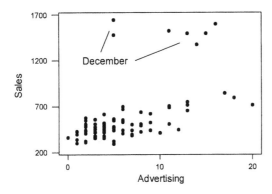

Exercise 10.5: Scatter plot

The results show that

```
PARAMETER   VARIABLE   NUM./   FACTOR   ORDER   CONS-       VALUE      STD       T
  LABEL       NAME     DENOM.                   TRAINT                 ERROR    VALUE

   1    CNST                     CNST     1       0    NONE   1478.9299  33.7541   43.81
   2    W1      TIME     NUM.     1       0    NONE       .4167    .1598    2.61
   3    W2      INDJAN   NUM.     1       0    NONE  -1140.0198  36.2740  -31.43
   4    W3      INDFEB   NUM.     1       0    NONE  -1149.4535  36.8407  -31.20
   5    W4      INDMAR   NUM.     1       0    NONE  -1064.2115  30.9192  -34.42
   6    W5      INDAPR   NUM.     1       0    NONE  -1034.2650  33.6153  -30.77
   7    W6      INDMAY   NUM.     1       0    NONE  -1043.4471  33.1970  -31.43
   8    W7      INDJUN   NUM.     1       0    NONE  -1045.3278  34.2302  -30.54
   9    W8      INDJUL   NUM.     1       0    NONE  -1112.7960  32.8056  -33.92
  10    W9      INDAUG   NUM.     1       0    NONE  -1046.2914  34.1008  -30.68
  11    W10     INDSEP   NUM.     1       0    NONE   -971.7751  32.3066  -30.08
```

74

```
12   W11     INDOCT   NUM.    1     0     NONE  -942.0039   30.9651 -30.42
13   W12     INDNOV   NUM.    1     0     NONE  -785.5350   34.2710 -22.92
14   W13      ADV     NUM.    1     0     NONE     2.0413    2.3240    .88
15   THETA   SALES    MA      1     1     NONE      .2522     .0999   2.52
```

```
EFFECTIVE NUMBER OF OBSERVATIONS . .       94
R-SQUARE . . . . . . . . . . . . .       0.966
RESIDUAL STANDARD ERROR. . . . . .  0.546787E+02

AUTOCORRELATIONS OF RESIDUALS

 1- 12     .00 -.03  .05 -.15 -.02 -.13 -.09 -.18  .01  .01  .07 -.09
 ST.E.     .10  .10  .10  .10  .11  .11  .11  .11  .11  .11  .11  .11

13- 24    -.04  .06 -.11  .04  .03  .01  .05  .03 -.06  .11  .02 -.02
 ST.E.     .11  .11  .11  .11  .11  .11  .11  .11  .11  .12  .12  .12
```

10.10 Scatter plots of ice cream consumption on price, family income, and temperature, and results of fitting the regression model

$\text{Cons}_t = \beta_0 + \beta_1 \text{Price}_t + \beta_2 \text{Inc}_t + \text{Temp}_t + \varepsilon_t$ are shown below. The Durbin-Watson statistic is much smaller than the desired value 2 and unacceptable. The small value of the Durbin-Watson statistic indicates positive lag one autocorrelation. The first six autocorrelations of the residuals are also shown. Especially the lag one autocorrelation $(r_1 = 0.32)$ is relatively large when compared to its standard error $1/\sqrt{30} = 0.18$.

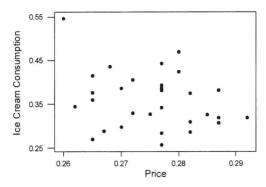

Exercise 10.10: Scatter plot

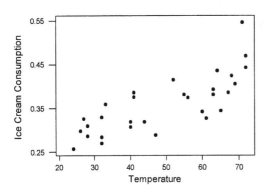

Exercise 10.10: Scatter plot

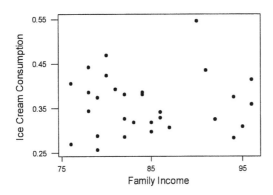

Exercise 10.10: Scatter plot

```
The regression equation is
Consumption = 0.197 - 1.04 Price + 0.00331 Income + 0.00346
Temperature

Predictor          Coef       SE Coef           T        P
Constant         0.1973        0.2702        0.73    0.472
Price           -1.0444        0.8344       -1.25    0.222
Income         0.003308      0.001171        2.82    0.009
Temperature   0.0034584     0.0004455        7.76    0.000


S = 0.03683     R-Sq = 71.9%      R-Sq(adj) = 68.7%

Analysis of Variance

Source             DF           SS           MS        F        P
Regression          3     0.090251     0.030084    22.17    0.000
Residual Error     26     0.035273     0.001357
Total              29     0.125523
```

```
Durbin-Watson statistic = 1.02
```

```
Autocorrelations of Residuals

   0.329772    0.036248    0.011063   -0.093395   -0.318641   -0.205802
```

The errors in this regression are not independent, and the error model needs to be revised. We consider two different error models: a first-order moving average and a first-order autoregressive error model. Note that in the regression with independent errors the coefficient for price is not significant. However, we keep this variable in the model as the significance may have been affected by the correlations in the errors. If it turns out that this coefficient is still insignificant, it can be removed at a later stage.

Estimation results for the two models are shown below. We use SCA to carry out the estimation. Alternatively, one can use EVIEWS. The residuals of the revised models are uncorrelated. The regression coefficients for income and temperature are significant (t-ratios exceed two). Income and temperature have positive regression coefficients; ice cream sales increase with increasing income and rising temperature. The coefficient of price is negative and not very significant (t-ratios of -1.77 and -1.18, respectively).

MA(1): $\text{Cons}_t = \beta_0 + \beta_1 \text{Price}_t + \beta_2 \text{Inc}_t + \beta_3 \text{Temp}_t + (1 - \theta B)a_t$

PARAMETER LABEL	VARIABLE NAME	NUM./ DENOM.	FACTOR	ORDER	CONS- TRAINT	VALUE	STD ERROR	T VALUE
1 B0		CNST	1	0	NONE	.3287	.2661	1.24
2 B1	PRICE	NUM.	1	0	NONE	-1.3886	.7829	-1.77
3 B2	INCOME	NUM.	1	0	NONE	.0029	.0014	2.15
4 B3	TEMP	NUM.	1	0	NONE	.0034	.0005	6.64
5 THETA	ICE	MA	1	1	NONE	-.5031	.1760	-2.86

```
EFFECTIVE NUMBER OF OBSERVATIONS . .        30
R-SQUARE . . . . . . . . . . . . . .      0.771
RESIDUAL STANDARD ERROR. . . . . . .  0.309303E-01

AUTOCORRELATIONS OF RESIDUALS

   1- 12    .02  .06 -.01  .02 -.30  .01 -.14 -.13 -.01 -.17 -.13  .07
   ST.E.    .18  .18  .18  .18  .18  .20  .20  .20  .21  .21  .21  .21

  13- 24    .32  .10  .02  .07  .13 -.15 -.04  .05 -.03 -.18  .01 -.23
   ST.E.    .21  .23  .23  .23  .23  .23  .24  .24  .24  .24  .24  .24
```

AR(1): $\text{Cons}_t = \beta_0 + \beta_1 \text{Price}_t + \beta_2 \text{Inc}_t + \beta_3 \text{Temp}_t + \dfrac{1}{(1 - \phi B)}a_t$

PARAMETER LABEL	VARIABLE NAME	NUM./ DENOM.	FACTOR	ORDER	CONS- TRAINT	VALUE	STD ERROR	T VALUE

```
    1   B0              CNST    1    0    NONE     .1495    .2697    .55
    2   B1    PRICE     NUM.    1    0    NONE    -.8889    .7532  -1.18
    3   B2    INCOME    NUM.    1    0    NONE     .0033    .0014   2.33
    4   B3    TEMP      NUM.    1    0    NONE     .0035    .0005   6.57
    5   PHI   ICE       D-AR    1    1    NONE     .4016    .1866   2.15

EFFECTIVE NUMBER OF OBSERVATIONS . .          29
R-SQUARE . . . . . . . . . . . . . .       0.790
RESIDUAL STANDARD ERROR. . . . . .   0.296282E-01

AUTOCORRELATIONS OF RESIDUALS

   1- 12    .09 -.11 -.02  .04 -.15  .08 -.10 -.09  .01 -.29 -.24  .09
   ST.E.    .19  .19  .19  .19  .19  .19  .19  .20  .20  .20  .21  .22

  13- 24    .38  .07 -.01  .00  .14 -.06  .02  .06  .03 -.09 -.12 -.20
   ST.E.    .22  .24  .24  .24  .24  .25  .25  .25  .25  .25  .25  .25
```

10.13 (a) Results of the regression
$$\text{FTEShares}_t = \beta_0 + \beta_1 \text{Car}\Pr\text{od}_{t-6} + \beta_2 \text{FTECom}_{t-7} + \varepsilon_t$$
are given below. The Durbin-Watson statistics is much smaller than the desired value 2 and is unacceptable. The small value of the Durbin-Watson statistic indicates positive lag one autocorrelation. The autocorrelation function of the residuals indicates significant autocorrelations, especially at lag 1 ($r_1 = 0.45$, compared to its standard error $1/\sqrt{22} = 0.13$). The extremely significant estimates for lagged car production and lagged commodity index are surprising, because results in the finance literature indicate that stock prices are best predicted by the current value of the stock, but not by other economic variables.

```
The regression equation is
FTEShare = 595 + 0.000514 CarLag6 - 5.54 ComLag7

55 cases used 7 cases contain missing values

Predictor        Coef       SE Coef          T         P
Constant       594.51         60.65       9.80     0.000
CarLag6    0.00051422    0.00003406      15.10     0.000
ComLag7       -5.5439        0.6727      -8.24     0.000

S = 25.06       R-Sq = 88.2%     R-Sq(adj) = 87.8%

Analysis of Variance

Source             DF           SS          MS        F        P
Regression          2       244274      122137   194.46    0.000
Residual Error     52        32661         628
Total              54       276935

Durbin-Watson statistic = 0.87
```

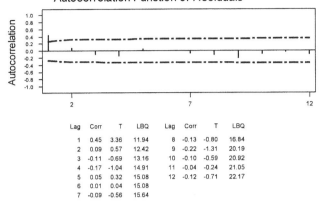

Autocorrelation Function of Residuals

Lag	Corr	T	LBQ	Lag	Corr	T	LBQ
1	0.45	3.36	11.94	8	-0.13	-0.80	16.84
2	0.09	0.57	12.42	9	-0.22	-1.31	20.19
3	-0.11	-0.69	13.16	10	-0.10	-0.59	20.92
4	-0.17	-1.04	14.91	11	-0.04	-0.24	21.05
5	0.05	0.32	15.08	12	-0.12	-0.71	22.17
6	0.01	0.04	15.08				
7	-0.09	-0.56	15.64				

(b) We consider the noisy random walk as a model for the errors, and fit the regression model

$$FTEShares_t = \beta_1 Car\,Prod_{t-6} + \beta_2 FTECom_{t-7} + \frac{1 - \theta B}{1 - B} a_t.$$

Because of the differencing operation, it is no longer possible to estimate the intercept β_0 of the earlier regression model.

```
--------------------------------------------------------------------
VARIABLE    TYPE OF     ORIGINAL       DIFFERENCING
            VARIABLE    OR CENTERED
                                           1
FTESHARE    RANDOM      ORIGINAL       (1-B  )
                                           1
  CAR       RANDOM      ORIGINAL       (1-B  )
                                           1
FTECOMM     RANDOM      ORIGINAL       (1-B  )
--------------------------------------------------------------------

PARAMETER   VARIABLE  NUM./  FACTOR  ORDER   CONS-      VALUE      STD       T
  LABEL       NAME    DENOM.                 TRAINT               ERROR    VALUE

   1    B1    CarProd   NUM.     1      6     NONE      .0001  .8107E-04   1.81
   2    B2    FTECom    NUM.     1      7     NONE     -.6884    1.1833    -.58
   3  THETA   FTEShares  MA      1      1     NONE     -.1468     .1417   -1.04


EFFECTIVE NUMBER OF OBSERVATIONS . .        54
R-SQUARE . . . . . . . . . . . . . .     0.951
RESIDUAL STANDARD ERROR. . . . . .   0.180416E+02

AUTOCORRELATIONS OF RESIDUALS

  1- 12    -.03 -.04  .00 -.12  .07  .02 -.15 -.00 -.32  .11 -.11 -.18
  ST.E.     .14  .14  .14  .14  .14  .14  .14  .14  .14  .15  .16  .16

 13- 24     .12 -.09  .03  .27 -.15  .09  .08  .10 -.01 -.17  .06 -.03
  ST.E.     .16  .16  .16  .16  .17  .17  .18  .18  .18  .18  .18  .18
```

(c) The estimate of θ is not much different from zero. We set it zero and estimate the parameters in the regression model with random walk errors

$$\text{FTEShares}_t = \beta_1 \text{Car Prod}_{t-6} + \beta_2 \text{FTECom}_{t-7} + \frac{1}{1-B} a_t \ .$$

This model is a regression of differences of the response on differences of the regressor variables,

$$\Delta \text{FTEShares}_t = \beta_1 \Delta \text{Car Prod}_{t-6} + \beta_2 \Delta \text{FTECom}_{t-7} + a_t \ .$$

The results given below show that there is no autocorrelation in the residuals. The model passes all diagnostic checks. The intercept and the regressors are not statistically significant (p-values of 0.085 and 0.51), implying that the model for the FTE share index is given by the random walk

$$\Delta \text{FTEShares}_t = \text{FTEShares}_t - \text{FTEShares}_{t-1} = a_t \ .$$

This result is expected. The finance literature shows that in efficient markets stock prices follow random walks and changes in stock prices are unrelated to economic variables. The "significant" regression in part (a) was spurious, implied by the incorrect model for the error terms; see the discussion of spurious regression in Section 10.2.

```
The regression equation is
DiffShare = 3.71 +0.000144 DiffCarLag6 - 0.79 DiffCommLag7

54 cases used 8 cases contain missing values

Predictor        Coef      SE Coef           T         P
Constant        3.712        2.547        1.46     0.151
DiffCarPr  0.00014414  0.00008218        1.75     0.085
DiffComm       -0.786        1.175       -0.67     0.507

S = 18.35       R-Sq = 6.1%       R-Sq(adj) = 2.4%

Analysis of Variance

Source            DF           SS          MS         F         P
Regression         2       1107.6       553.8      1.64     0.203
Residual Error    51      17179.0       336.8
Total             53      18286.5

Durbin-Watson statistic = 1.72

Autocorrelations of residuals

    0.100215  -0.050381  -0.024561  -0.117140   0.070461   0.023664
   -0.147233  -0.070849  -0.317866   0.050094  -0.118589  -0.173089
```

CHAPTER 11

A note on computing with MINITAB (Version 14):

The **Minitab** software is used for fitting the logistic regression models in Chapter 11. Alternatively, one can use the SAS PROC GENMOD procedure; see the explanation in Chapter 12 of this solutions manual.

Minitab works like a spreadsheet program. We enter the data into the various columns of the spreadsheet and use the tabs: Stat > Regression > Binary logistic regression. We need to specify the response; either a column of zeros and ones if we work with individual cases, or the number of successes and the number of trials for each constellation if we work with aggregated data. We need to write out the model in model format. We can declare variables as factors – then Minitab will automatically create the needed indicator variables and test for factor effects. We can store the results (fitted values, residuals, …) in unused columns of the worksheet. All diagnostic graphs discussed in Chapter 11 of the book are available in Minitab.

Options for various links (logit, probit, and complementary log-log links), starting values, maximum number of iterations, and number of classes in the Hosmer-Lemeshow test are available. Many other options are available. See the Minitab on-line help for detailed discussion and examples.

11.1 Time series graphs of weekly proportions of long fibers are given below. 7-term moving averages, $MA_t = (y_{t-2} + y_{t-1} + y_t + y_{t+1} + y_{t+2})/7$, are added to these graphs. Moving averages amplify the trend component in a time series graph of noisy observations. The proportions of long fibers increase during the second half of the year.

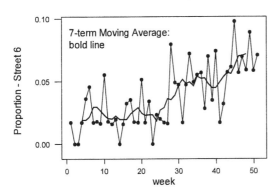

Exercise 11.1: Time series plot - Street 6

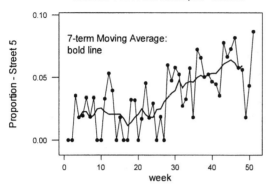

Exercise 11.1: Time series plot - Street 5

For each street (machine) separately, we construct scatter plots of the proportions of long fibers against stretch reduction, total throughput, and the type of process. The proportions of long fibers decrease with increased stretch reduction. The proportion of long fibers is larger under process 2.

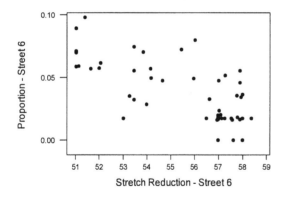

Exercise 11.1: Scatter plot - Street 6

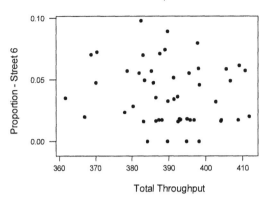

Exercise 11.1: Scatter plot - Street 6

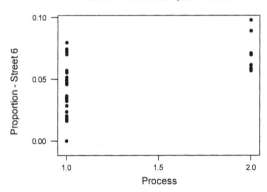

Exercise 11.1: Scatter plot - Street 6

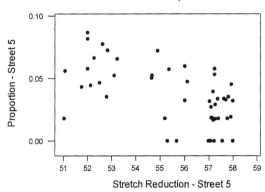

Exercise 11.1: Scatter plot - Street 5

83

Exercise 11.1: Scatter plot - Street 5

Logistic regression models for machine (street) 6:

Results for the following three logistic regression models are given below:

- model with stretch reduction, throughput, and process
- model with stretch reduction and throughput
- model with stretch reduction only

The total throughput and the type of process are insignificant. Stretch reduction remains as the only significant variable. An increase in the stretch reduction of one unit (percent) changes the odds for long fibers by a (multiplicative) factor of 0.85. That is, an increase in the stretch reduction of one unit (percent) reduces the odds for the occurrence of long fibers by 15 percent. Or, to say this differently: A small stretch reduction increases the odds for quality problems.

The proportion of long fibers π can be obtained from

$$\pi(x) = \frac{\exp(\beta_0 + \beta_1 x)}{1 + \exp(\beta_0 + \beta_1 x)} = \frac{\exp(5.928 - 0.1662x)}{1 + \exp(5.928 - 0.1662x)}$$

For stretch reduction x = 52, $\pi(x = 52) = \dfrac{\exp(5.928 - 0.1662(52))}{1 + \exp(5.928 - 0.1662(52))} = 0.062$

For stretch reduction x = 53, $\pi(x = 53) = \dfrac{\exp(5.928 - 0.1662(53))}{1 + \exp(5.928 - 0.1662(53))} = 0.053$

….

For stretch reduction x = 57, $\pi(x = 57) = \dfrac{\exp(5.928 - 0.1662(57))}{1 + \exp(5.928 - 0.1662(57))} = 0.028$

We have superimposed the fitted values (proportions of long fibers) in the scatter plot of the proportion of long fibers against stretch reduction (street 6). The main features of the scatter plot are well represented by the fitted model.

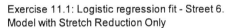

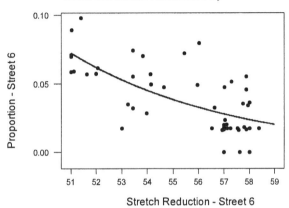

In this problem there are few exact replicates of the explanatory variable, stretch reduction. Minitab uses the approach by Hosmer and Lemeshow to group the cases on the basis of the estimated probabilities $\hat{\pi}_i = \hat{\pi}(x_i)$. It ranks the estimated probabilities from the smallest to the largest, and uses this ranking to break the cases into $g = 10$ groups of equal size. For each group k, $k = 1, 2, ..., g$, it calculates the number of successes o_k and the number of failures $n_k - o_k$ that are associated with the n_k cases in the group. The observed frequencies are compared with the expected frequencies $n_k \bar{\pi}_k$ and $n_k (1 - \bar{\pi}_k)$, where $\bar{\pi}_k = \dfrac{\sum_{i \in \text{Group} k} \hat{\pi}_i}{n_k}$ is the average estimated success probability in the kth group. The Pearson chi-square statistic is calculated from the resulting $2 \times g$ table, and

$$HL = \sum_{k=1}^{g} \frac{[o_k - n_k \bar{\pi}_k]^2}{n_k \bar{\pi}_k (1 - \bar{\pi}_k)}$$

is referred to as the Hosmer-Lemeshow statistic. Hosmer and Lemeshow show that the distribution of HL is well approximated by a chi-square distribution with $g - 2$ degrees of freedom. Large values of the Hosmer-Lemeshow statistic indicate lack of

85

fit. In our problem the Hosmer-Lemeshow statistic is $HL = 6.938$. It is quite small when compared to the 95th percentile of chi-square distribution with $10 - 2 = 8$ degrees of freedom (15.51). The associated large probability value, 0.435, confirms that the model gives a very adequate representation of the data.

The Pearson residual for each of the 52 weeks is calculated from the equation

$$r_i = r(y_i, \hat{\pi}_i) = \frac{y_i - n_i\hat{\pi}_i}{\sqrt{n_i\hat{\pi}_i(1-\hat{\pi}_i)}},$$ where y_i and n_i are the number of occurrences and the

total number of trials in week i, and where

$$\hat{\pi}_i = \hat{\pi}(x_i) = \frac{\exp(\hat{\beta}_0 + \hat{\beta}_1 x_i)}{1+\exp(\hat{\beta}_0 + \hat{\beta}_1 x_i)} = \frac{\exp(5.928 - 0.1662x_i)}{1+\exp(5.928 - 0.1662x_i)}$$ is the implied success

probability. The autocorrelations for the first six lags are given by

```
0.00, 0.06, -0.22, 0.05, 0.16, -0.02.
```

Comparing these to their approximate standard error, $1/\sqrt{52} = 0.14$, indicates no serial correlation among the residuals.

A note on residuals and fitted values: Minitab stores the residuals and the diagnostic measures for each constellation, and the constellations change with different model specifications. When estimating the logistic regression on stretch reduction alone, there are data for 51 weeks, but there are only 43 different stretch constellations. For three weeks the stretch reduction on street 6 is 51.000, for four weeks it is 57.000, and for four weeks it is 58.000. Minitab aggregates the information and supplies vectors of fitted values and residuals for the 43 constellations. This is fine as far as the usual diagnostic checks are concerned, but it causes difficulties if one wants to calculate the autocorrelations of the residuals where time order is of importance. One cannot compute the autocorrelations of weekly residuals from the vector of the aggregated residuals.
One must first compute the residuals for each week. This can be done by using the weekly frequencies (number of successes and number of trials) and the event probabilities at the constellations (note that these are stored by Minitab).
Alternatively, one can "trick" the program by adding small numbers to the replicates of stretch to make them slightly different (say 51.000, 51.001, and 51.003 for the three weeks with identical stretch reduction 51.000; etc). Then Minitab will treat them as separate constellations and will give you the vector of the 51 weekly residuals automatically.

Model with stretch reduction, throughput, and process:

```
Link Function:  Logit

Response  Information
```

```
Variable  Value              Count
Positive6 Success              139
          Failure             3225
samples6  Total               3364
```

Logistic Regression Table

Predictor	Coef	SE Coef	Z	P	Odds Ratio	95% CI Lower	95% CI Upper
Constant	8.854	4.044	2.19	0.029			
stretch6	-0.16900	0.06532	-2.59	0.010	0.84	0.74	0.96
throughput	-0.007084	0.008151	-0.87	0.385	0.99	0.98	1.01
process6	-0.0062	0.3606	-0.02	0.986	0.99	0.49	2.02

```
Log-Likelihood = -566.074
Test that all slopes are zero: G = 25.849, DF = 3, P-Value = 0.000
```

Goodness-of-Fit Tests

Method	Chi-Square	DF	P
Pearson	29.837	47	0.976
Deviance	32.323	47	0.949
Hosmer-Lemeshow	2.053	8	0.979

Table of Observed and Expected Frequencies:
(See Hosmer-Lemeshow Test for the Pearson Chi-Square Statistic)

Value	1	2	3	4	5	6	7	8	9	10	Total
Success											
Obs	8	10	7	8	13	20	17	19	26	11	139
Exp	7.7	9.3	9.1	10.0	10.5	17.7	17.2	20.0	25.7	11.8	
Failure											
Obs	335	378	337	348	331	365	321	323	342	145	3225
Exp	335.3	378.7	334.9	346.0	333.5	367.3	320.8	322.0	342.3	144.2	
Total	343	388	344	356	344	385	338	342	368	156	3364

Model with stretch reduction and throughput:

```
Link Function:  Logit

Response  Information

Variable  Value              Count
Positive6 Success              139
          Failure             3225
samples6  Total               3364
```

Logistic Regression Table

Predictor	Coef	SE Coef	Z	P	Odds Ratio	95% CI Lower	95% CI Upper
Constant	8.818	3.461	2.55	0.011			
stretch6	-0.16805	0.03391	-4.96	0.000	0.85	0.79	0.90
throughput	-0.007146	0.007285	-0.98	0.327	0.99	0.98	1.01

```
Log-Likelihood = -566.075
Test that all slopes are zero: G = 25.849, DF = 2, P-Value = 0.000

Goodness-of-Fit Tests

Method              Chi-Square    DF      P
Pearson                29.840     48  0.982
Deviance               32.324     48  0.960
Hosmer-Lemeshow         2.063      8  0.979

Table of Observed and Expected Frequencies:
(See Hosmer-Lemeshow Test for the Pearson Chi-Square Statistic)

                              Group
Value    1     2     3     4     5     6     7     8     9    10   Total
Success
  Obs     8    10     7     8    13    20    17    19    26    11    139
  Exp   7.7   9.3   9.1  10.0  10.5  17.6  17.2  20.0  25.7  11.8
Failure
  Obs   335   378   337   348   331   365   321   323   342   145   3225
  Exp 335.3 378.7 334.9 346.0 333.5 367.4 320.8 322.0 342.3 144.2

 Total  343   388   344   356   344   385   338   342   368   156   3364
```

Model with stretch reduction only:

```
Link Function:  Logit

Response  Information

Variable Value        Count
Positive6 Success       139
          Failure      3225
samples6 Total         3364

Logistic Regression Table
                                              Odds      95% CI
Predictor       Coef    SE Coef      Z    P   Ratio   Lower   Upper
Constant       5.928     1.818    3.26 0.001
stretch6    -0.16619   0.03359   -4.95 0.000   0.85    0.79    0.90

Log-Likelihood = -566.554
Test that all slopes are zero: G = 24.891, DF = 1, P-Value = 0.000

Goodness-of-Fit Tests

Method              Chi-Square    DF      P
Pearson                27.801     41  0.943
Deviance               26.951     41  0.955
Hosmer-Lemeshow         6.938      7  0.435

Table of Observed and Expected Frequencies:
(See Hosmer-Lemeshow Test for the Pearson Chi-Square Statistic)
```

Value	1	2	3	4	5	6	7	8	9	Total
Success										
Obs	9	8	11	6	20	21	17	25	22	139
Exp	9.2	8.8	10.4	11.1	13.1	18.5	19.4	26.0	22.4	
Failure										
Obs	379	345	373	389	367	383	340	363	286	3225
Exp	378.8	344.2	373.6	383.9	373.9	385.5	337.6	362.0	285.6	
Total	388	353	384	395	387	404	357	388	308	3364

Logistic regression models for machine (street) 5:

Results for the following two logistic regressions models are given below:

- model with stretch reduction and throughput
- model with stretch reduction only

Process does not enter here, as machine 5 operates under one production process.

Total throughput is insignificant. Stretch reduction remains as the only significant variable. An increase in the stretch reduction of one unit (percent) changes the odds for long fibers by a (multiplicative) factor of 0.85. That is, an increase in the stretch reduction of one unit (percent) reduces the odds for long fibers by 15 percent. Note that the odds-ratios for stretch reduction are the same on both streets.

The proportion of long fibers π can be obtained from

$$\pi(x) = \frac{\exp(\beta_0 + \beta_1 x)}{1 + \exp(\beta_0 + \beta_1 x)} = \frac{\exp(6.006 - 0.1681x)}{1 + \exp(6.006 - 0.1681x)}$$

For stretch reduction x = 52, $\pi(x = 52) = \dfrac{\exp(6.006 - 0.1681(52))}{1 + \exp(6.006 - 0.1681(52))} = 0.061$

For stretch reduction x = 53, $\pi(x = 53) = \dfrac{\exp(6.006 - 0.1681(53))}{1 + \exp(6.006 - 0.1681(53))} = 0.052$

....

For stretch reduction x = 57, $\pi(x = 57) = \dfrac{\exp(6.006 - 0.1681(57))}{1 + \exp(6.006 - 0.1681(57))} = 0.027$

We have superimposed the fitted proportions of long fibers in the scatter plot of the proportion of long fibers against stretch reduction (street 5). The main features of the scatter plot are well represented by the fitted model.
The Hosmer-Lemeshow statistic is $HL = 5.146$. It is quite small when compared with the 95th percentile of chi-square distribution with $10 - 2 = 8$ degrees of freedom

(15.51). The associated large probability value, 0.742, confirms that the model leads to a very adequate representation of the data.

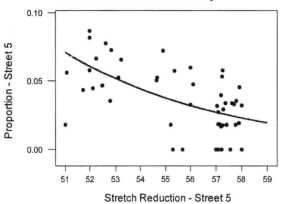

Exercise 11.1: Logistic regression fit - Street 5
Model with Stretch Reduction Only

Model with stretch reduction and throughput:

```
Link Function:  Logit

Response  Information

Variable  Value          Count
Positive5 Success          119
          Failure         3014
samples5  Total           3133


Logistic Regression Table
                                                Odds      95% CI
Predictor        Coef    SE Coef      Z     P   Ratio  Lower  Upper
Constant        9.888     3.934    2.51 0.012
stretch5       -0.17408   0.03979  -4.38 0.000   0.84   0.78   0.91
throughput     -0.009107  0.007761 -1.17 0.241   0.99   0.98   1.01

Log-Likelihood = -496.083
Test that all slopes are zero: G = 19.664, DF = 2, P-Value = 0.000

Goodness-of-Fit Tests

Method               Chi-Square    DF     P
Pearson                 36.191     48   0.895
Deviance                47.990     48   0.473
Hosmer-Lemeshow          5.904      7   0.551
Table of Observed and Expected Frequencies:
(See Hosmer-Lemeshow Test for the Pearson Chi-Square Statistic)
```

```
                                   Group
Value       1     2     3     4     5     6     7     8     9    Total
Success
  Obs        7    10     4    11     8    18    21    21    19     119
  Exp      7.3   8.2   8.5   9.3  10.4  14.5  18.3  19.5  23.0
Failure
  Obs      326   336   328   335   341   361   348   316   323    3014
  Exp    325.7 337.8 323.5 336.7 338.6 364.5 350.7 317.5 319.0

  Total    333   346   332   346   349   379   369   337   342    3133
```

Model with stretch reduction only:

```
Link Function:  Logit

Response  Information

Variable  Value        Count
Positive5 Success        119
          Failure       3014
samples5  Total         3133

Logistic Regression Table
                                                  Odds      95% CI
Predictor       Coef    SE Coef       Z     P    Ratio   Lower  Upper
Constant       6.006     2.138     2.81 0.005
stretch5     -0.16807   0.03917    -4.29 0.000    0.85    0.78   0.91

Log-Likelihood = -496.765
Test that all slopes are zero: G = 18.300, DF = 1, P-Value = 0.000

Goodness-of-Fit Tests

Method               Chi-Square   DF      P
Pearson                 33.514    43   0.850
Deviance                44.313    43   0.416
Hosmer-Lemeshow          5.146     8   0.742

Table of Observed and Expected Frequencies:
(See Hosmer-Lemeshow Test for the Pearson Chi-Square Statistic)

                                   Group
Value       1     2     3     4     5     6     7     8     9    10   Total
Success
  Obs        8     8    10     6    12    11    21    20    22     1     119
  Exp      7.9   8.5   8.5   9.2  11.0  11.4  16.6  20.2  21.9   3.9
Failure
  Obs      329   335   314   336   355   304   330   334   323    54    3014
  Exp    329.1 334.5 315.5 332.8 356.0 303.6 334.4 333.8 323.1  51.1

  Total    337   343   324   342   367   315   351   354   345    55    3133
```

11.3 The information can be arranged as a factorial, with the number of affected workers among the total number of workers in each group as the response variable. The 72 groups of the factorial arrangement are formed by all possible level combinations of the five explanatory variables: 3 (Dust) x 2 (Race) x 2 (Sex) x 2 (Smoking) x 3 (Employment). Seven of the 72 categories are empty and are ignored in our analysis. We use the binary logistic regression function in MINITAB, specifying the number of successes and the number of trials, and entering the explanatory variables as (categorical) factors. MINITAB creates the appropriate indicators for the factors automatically.

Model with all five factors:

```
Link Function:  Logit

Response  Information

Variable  Value        Count
Yes       Success        165
          Failure       5254
Number    Total         5419

   65 cases were used
    7 cases contained missing values

Logistic Regression Table
                                                Odds      95% CI
Predictor      Coef    SE Coef      Z     P    Ratio   Lower   Upper
Constant     -1.9452    0.2334   -8.33 0.000
Dust
  2          -2.5799    0.2921   -8.83 0.000    0.08    0.04    0.13
  3          -2.7306    0.2153  -12.68 0.000    0.07    0.04    0.10
Race
  2           0.1163    0.2072    0.56 0.574    1.12    0.75    1.69
Sex
  2           0.1239    0.2288    0.54 0.588    1.13    0.72    1.77
Smoking
  2          -0.6413    0.1944   -3.30 0.001    0.53    0.36    0.77
Employ
  2           0.5641    0.2617    2.16 0.031    1.76    1.05    2.94
  3           0.7531    0.2161    3.48 0.000    2.12    1.39    3.24

Log-Likelihood = -598.968
Test that all slopes are zero: G = 279.256, DF = 7, P-Value = 0.000

Goodness-of-Fit Tests

Method            Chi-Square   DF    P
Pearson              37.934    57  0.976
Deviance             43.271    57  0.910
```

The test statistic for testing the overall significance of the regression (in equation (11.25)) is given by G = 279.256. Its sampling distribution (under the null hypotheses that none of the regressors have an influence on the response) is chi-square with 7 degrees of freedom. The test statistic G = 279.256 is huge compared to the percentiles from that distribution, and its associated probability value is tiny (p value < 0.0001).

Hence the regressor variables (all or a subset) have a significant impact on the occurrence of byssinosis.

Race and Sex (both at two levels) have no significant effects. One can see this from the odds-ratios (they are roughly one), their t-ratios (Z-scores) and the associated probability values. The probability values for Race and Sex exceed the usual cutoff 0.05. The insignificance of the effects is also expressed by the confidence intervals of the odds-ratios; the confidence intervals cover one (indicating even odds).

The dustiness of the workplace, the smoking history, and the length of employment matter; the probability values of the estimated coefficients are smaller than 0.05, and the confidence intervals of the resulting odds-ratios do not cover the value one.

The deviance (in equation (11.26)) and the Pearson statistic (in equation (11.31)) compare the fit of the parameterized model (here with $8 = 7 + 1$ (for constant) parameters) with the fit of the saturated model where each constellation of the explanatory variables is allowed its own distinct success probability. Here there are 65 $= 72 - 7$ constellations as seven cells are empty. The deviance is $D = 37.9$ and the Pearson statistic is $\chi^2 = 43.3$. Large values of these statistics indicate model inadequacy; the appropriate reference distribution is chi-square with $65 - 8 = 57$ degrees of freedom. The deviance and the Pearson statistic are smaller than the critical percentile (the 95th percentile is 75.62), implying that the probability values are considerably larger than 0.05. Hence there is no reason to question the adequacy of the model.

Here the deviance and the Pearson chi-square statistics are useful measures of (lack of) fit, as we have replicate observations at each configuration of the explanatory variable(s). In this example there is no reason to consider the Hosmer-Lemeshow statistic which becomes useful if we don't have replicate observations (as is often the case with continuous covariates).

The next steps in the analysis remove the insignificant regressors, sex and race. Because of possible multicollinearity it is always safer to this one step at a time. We first omit race as this variable has the smaller insignificant t-ratio (or, equivalently, the larger probability value). The output of the simplified model is given below:

Model without race:

```
Link Function:   Logit

Response  Information

Variable  Value           Count
Yes       Success           165
          Failure          5254
Number    Total            5419
```

```
    65 cases were used
     7 cases contained missing values

Logistic Regression Table
                                                 Odds      95% CI
Predictor     Coef    SE Coef      Z      P     Ratio   Lower   Upper
Constant    -1.8483    0.1549  -11.93 0.000
Dust
  2          -2.6118    0.2864   -9.12 0.000    0.07    0.04    0.13
  3          -2.7623    0.2079  -13.29 0.000    0.06    0.04    0.09
Sex
  2           0.1247    0.2286    0.55 0.586    1.13    0.72    1.77
Smoking
  2          -0.6411    0.1944   -3.30 0.001    0.53    0.36    0.77
Employ
  2           0.5238    0.2512    2.08 0.037    1.69    1.03    2.76
  3           0.6904    0.1844    3.74 0.000    1.99    1.39    2.86

Log-Likelihood = -599.126
Test that all slopes are zero: G = 278.940, DF = 6, P-Value = 0.000

Goodness-of-Fit Tests

Method            Chi-Square    DF      P
Pearson             28.316      27    0.395
Deviance            29.716      27    0.327
```

The factor sex is insignificant (t-ratio 0.55, and probability value 0.59), and is omitted in the next model.

Model without race and sex:

```
Link Function:  Logit

Response  Information

Variable  Value       Count
Yes       Success       165
          Failure      5254
Number    Total        5419

    65 cases were used
     7 cases contained missing values

Logistic Regression Table
                                                 Odds      95% CI
Predictor     Coef    SE Coef      Z      P     Ratio   Lower   Upper
Constant    -1.8336    0.1525  -12.03 0.000
Dust
  2          -2.5493    0.2614   -9.75 0.000    0.08    0.05    0.13
  3          -2.7175    0.1898  -14.31 0.000    0.07    0.05    0.10
Smoking
  2          -0.6210    0.1908   -3.26 0.001    0.54    0.37    0.78
Employ
  2           0.5060    0.2490    2.03 0.042    1.66    1.02    2.70
  3           0.6728    0.1813    3.71 0.000    1.96    1.37    2.80

Log-Likelihood = -599.274
```

```
Test that all slopes are zero: G = 278.645, DF = 5, P-Value = 0.000

Goodness-of-Fit Tests

Method             Chi-Square   DF    P
Pearson               13.570    12  0.329
Deviance              12.094    12  0.438
```

No other variables can be omitted. Smoking is an important contributor to byssinosis. For a non-smoker the odds of contracting byssinosis are 0.54 the odds of a smoker. Everything else equal, not smoking reduces the odds of contracting byssinosis by 46 percent.

The length of employment in the cotton industry matters. The odds that a worker with 10 to 20 years employment contracts byssinosis are 1.66 times the odds of a worker with less than ten years in the industry. The odds for a worker with more than 20 years are twice (1.96) the odds of a worker with less than ten years in the industry.

Dustiness of the workplace clearly matters. The odds of contracting byssinosis at workplaces with medium and low levels of dustiness are considerably smaller than the odds for workplaces with a high level of dustiness (they are 0.08 and 0.07 times the odds of workplaces with high level of dustiness).

Next, we explore whether it is necessary to include interactions. The model with the three factors - smoking, length of employment, and dustiness of the workplace - and all two-factor interactions is given below.

Model with two-factor interactions:

```
Link Function:  Logit

Response  Information

Variable  Value          Count
Yes       Success          165
          Failure         5254
Number    Total           5419

Factor Information

Factor    Levels Values
Dust         3 1 2 3
Smoking      2 1 2
Employ L     3 1 2 3

    65 cases were used
     7 cases contained missing values

Logistic Regression Table
                                               Odds       95% CI
Predictor           Coef    SE Coef       Z     P  Ratio   Lower    Upper
Constant         -1.9545     0.1922  -10.17 0.000
Dust
```

```
2                  -2.7064    0.4775    -5.67 0.000      0.07     0.03     0.17
3                  -2.4646    0.3274    -7.53 0.000      0.09     0.04     0.16
Smoking
2                  -0.7242    0.3516    -2.06 0.039      0.48     0.24     0.97
Employ
2                   0.8287    0.3324     2.49 0.013      2.29     1.19     4.39
3                   0.9904    0.2551     3.88 0.000      2.69     1.63     4.44
Dust*Smoking
2*2                 1.1956    0.5501     2.17 0.030      3.31     1.12     9.72
3*2                 0.4546    0.4375     1.04 0.299      1.58     0.67     3.71
Dust*Employ
2*2                -0.1908    0.7751    -0.25 0.806      0.83     0.18     3.78
2*3                -0.5094    0.5881    -0.87 0.386      0.60     0.19     1.90
3*2                -1.0915    0.6432    -1.70 0.090      0.34     0.10     1.18
3*3                -0.4572    0.4103    -1.11 0.265      0.63     0.28     1.41
Smoking*Employ
2*2                -0.0556    0.6162    -0.09 0.928      0.95     0.28     3.16
2*3                -0.4911    0.4183    -1.17 0.240      0.61     0.27     1.39

Tests for terms with more than 1 degree of freedom

Term              Chi-Square    DF      P
Dust                 73.005     2   0.000
Employ               16.025     2   0.000
Dust*Smoking          4.863     2   0.088
Dust*Employ           3.712     4   0.446
Smoking*Employ        1.473     2   0.479

Log-Likelihood = -593.735
Test that all slopes are zero: G = 289.723, DF = 13, P-Value = 0.000

Goodness-of-Fit Tests
Method          Chi-Square    DF      P
Pearson              1.005     4   0.909
Deviance             1.016     4   0.907
```

The interactions between dust and employment length and between smoking history and employment length matter little, and are omitted from the model at the next step. The chi-square tests for the Dust*EmployLength interaction is 3.712 with probability value 0.446, and the Smoking*EmployLength interaction is 1.473 with probability value 0.479. These chi-square tests compare the full model with the model that restricts the interactions under consideration to zero.

Fitting the simpler model with the three factors smoking, length of employment, and dustiness of the workplace and the remaining 2-factor interaction between dust and smoking is shown below.

Model with the dustiness by smoking interaction:

```
Link Function:  Logit
Response  Information

Variable  Value         Count
Yes       Success         165
          Failure        5254
Number    Total          5419
```

96

```
Factor Information

Factor     Levels Values
Dust          3 1 2 3
Smoking       2 1 2
Employ L      3 1 2 3

    65 cases were used
     7 cases contained missing values


Logistic Regression Table
                                                  Odds      95% CI
Predictor       Coef    SE Coef      Z     P     Ratio   Lower   Upper
Constant      -1.7573    0.1555  -11.30 0.000
Dust
  2            -2.9576    0.3565   -8.30 0.000    0.05    0.03    0.10
  3            -2.8325    0.2230  -12.70 0.000    0.06    0.04    0.09
Smoking
  2            -0.9573    0.2751   -3.48 0.001    0.38    0.22    0.66
Employ
  2             0.4990    0.2499    2.00 0.046    1.65    1.01    2.69
  3             0.6638    0.1819    3.65 0.000    1.94    1.36    2.77
Dust*Smoking
  2*2           1.1807    0.5490    2.15 0.031    3.26    1.11    9.55
  3*2           0.4864    0.4338    1.12 0.262    1.63    0.69    3.81


Tests for terms with more than 1 degree of freedom

Term          Chi-Square    DF      P
Dust             198.232     2   0.000
Employ            13.717     2   0.001
Dust*Smoking       4.840     2   0.089

Log-Likelihood = -596.848
Test that all slopes are zero: G = 283.496, DF = 7, P-Value = 0.000

Goodness-of-Fit Tests

Method        Chi-Square    DF      P
Pearson            7.289    10   0.698
Deviance           7.243    10   0.702
```

We illustrate in detail how one can test whether the interaction is significant. Comparing the log-likelihood = -596.848 of the full model with the log-likelihood of the restricted model (model without the interaction; log-likelihood = -599.274) leads to the log-likelihood ratio test statistic $2(-596.848-(-599.274)) = 4.84$. Relating this statistic to a chi-square distribution with 2 degrees of freedom leads to the probability value $P(\chi^2(2) \geq 4.84) = 0.089$. Note that the test-statistic (4.84) and the probability value (0.089) are given in the previous computer output. Since the probability value is larger than 0.05, we conclude that the interaction is not significant. Of course, at the ten percent significance level one would conclude that there is a smoking by dustiness interaction effect on the odds of contracting byssinosis. While there is some evidence of an interaction, the evidence is certainly not very strong.

How would one interpret the coefficients and the odds-ratios in the interaction component? One can write out the logistic regression model with the interaction terms and look at the odds for fixed levels of dustiness of the workplace.

(i) Comparing the odds for a non-smoker at a high-level dusty workplace (dust level 1), exp(constant - 0.9573), to those of a smoker at a high-level dusty workplace, exp(constant), leads to the odds-ratio exp(-0.9573) = 0.38. At a dusty workplace, nonsmoking reduces the odds of contracting byssinosis by 62 percent.

(ii) The odds-ratio for a non-smoker at a medium-level dusty workplace (dust level 2) is 0.38exp(1.1807) = (0.38)(3.26) = 1.25. At a medium-level dusty workplace the odds of contracting byssinosis for smokers and non-smokers are about the same. At medium-level dusty workplaces the smoking history has little influence on the odds of contracting the disease.

(iii) The odds-ratio for a non-smoker at a low-level dusty workplace (dust level 3) is 0.38exp(0.4864) = (0.38)(1.63) = 0.62. However, note the confidence interval for the interaction effect for (non)smoking and low dustiness (level 3) is quite wide (extending from 0.69 to 3.81) making the interpretation for low-level dustiness quite uncertain. The odds of contracting byssinosis for smokers and non-smokers may in fact be the same.

In summary, nonsmoking reduces the odds of contracting byssinosis, and the reduction is largest in very dusty workplaces.

CHAPTER 12

A note on computing with SAS (Version 9):

The **SAS GENMOD** procedure is used for fitting the Poisson regression models of Chapter 12. This procedure is very general. It can also be used for the logistic regression models in Chapter 11, as well as most generalized linear models.

SAS works slightly different than the previously considered spreadsheet programs Minitab, SPSS, or EXCEL. In SAS one needs to write out a line code. The line code gets entered into a program editor, and is executed by clicking the SAS "run" and "submit" tabs. Here we list an example of the line code, with a detailed discussion of important options. Many more options are available, and they can be reviewed by looking at the on-line help pages within SAS.

We list the input for Exercise 12.1:

```
data exer12n1;
            specifies the file name for data set
input type year period ms nudamage;
            specifies the input variables
lnms=log(ms);
            specifies a transformation; here the natural log transformation
datalines;
1     1     1     127     0
1     1     2     63      0
1     2     1     1095    3
1     2     2     1095    4
1     3     1     1512    6
1     3     2     3353    18
1     4     2     2244    11
2     1     1     44882   39
2     1     2     17176   29
2     2     1     28609   58
2     2     2     20370   53
2     3     1     7064    12
2     3     2     13099   44
2     4     2     7117    18
3     1     1     1179    1
3     1     2     552     1
3     2     1     781     0
3     2     2     676     1
3     3     1     783     6
3     3     2     1948    2
```

```
3     4     2     274     1
4     1     1     251     0
4     1     2     105     0
4     2     1     288     0
4     2     2     192     0
4     3     1     349     2
4     3     2     1208    11
4     4     2     2051    4
5     1     1     45      0
5     2     1     789     7
5     2     2     437     7
5     3     1     1157    5
5     3     2     2161    12
5     4     2     542     1
;
proc genmod data=exer12n1;
```
PROC GENMOD is called
```
class type / param=ref ref=first;
class year / param=ref ref=first;
class period / param=ref ref=first;
```
specifies that type, year, and period are class (factor) variables; SAS creates the appropriate indicator variables automatically. The first numeric value is taken as the base for comparisons.
```
model nudamage=type year period lnms / d=poisson obstats
covb corrb lrci type3;
```
Here the model gets specified. The response is nudamage. The first three variables on the right hand side of the equal sign are factors. The last variable (lnms) is a covariate (not a factor). Options are listed after the slash.

d=Poisson: Poisson link.

Covb, Corrb: Covariance and correlation matrices of the parameter estimates are displayed.

Obstats: results in detailed output (fitted values, residuals, ...)

Lrci requests that two-sided confidence intervals for all model parameters are computed based on the profile likelihood function. This is sometimes called the partially maximized likelihood function. Two-sided Wald confidence intervals are calculated, if lrci is not specified.

Likelihood ratio-based confidence intervals, also known as profile likelihood confidence intervals, of parameter estimates in generalized linear models can be explained as follows. Suppose that the parameter vector is $\beta = (\beta_0, \beta_1, ..., \beta_p)'$ and one wants a confidence interval for β_i. The profile likelihood function for β_i is defined as $l*(\beta_i) = \max_{\tilde{\beta}} l(\beta)$, where $\tilde{\beta}$ is the vector β with the ith element fixed at β_i and $l = l(\beta)$ is the log likelihood function. Let $l = l(\hat{\beta})$ be the log likelihood evaluated at the maximum likelihood estimate $\hat{\beta}$. Under the assumption that β_i is the true parameter value, $2(l - l*(\beta_i))$ has a limiting chi-square distribution with one degree of freedom. A $100(1 - \alpha)$ percent confidence interval for β_i is

$$\{\beta_i : l*(\beta_i) \geq l - 0.5\chi^2(1 - \alpha;1)\}$$

where $\chi^2(1 - \alpha;1)$ is the $100(1 - \alpha)$ percentile of the chi-square distribution with one degree of freedom. The endpoints of the confidence interval can be found by solving numerically for values of β_i that satisfy the equality in the preceding relation.

Type 3: requests that statistics for Type 3 contrasts be computed for each class variable (factor) specified in the MODEL statement. This means that likelihood-ratio tests are calculated for the contrasts of the class variables. Type 3 means that these are partial tests, comparing the full model with the restricted model that lacks the indicated class variable (factor).

OFFSET = lnms: specifies a variable in the input data set (here lnms) to be used as an offset variable. This variable cannot be a CLASS variable. In our example it seems reasonable to suppose that the number of damage incidents is directly proportional to MS, the months of service, and one can expect that the coefficient in the Poisson regression model that corresponds to ln(MS) is one. OFFSET = lnms restricts this parameter to one.

Scale = deviance: Overdispersion is a phenomenon that sometimes occurs in data that are modeled with the Poisson (and also binomial - see Chapter 11) distributions. If the estimate of dispersion after fitting, as measured by the deviance or Pearson's chi-square divided by the degrees of freedom, is not near 1, then the data may be overdispersed if the dispersion estimate is greater than 1, or underdispersed if the dispersion estimate is less than 1. A simple way to model this situation is to allow the variance function of the Poisson distribution to have a multiplicative overdispersion factor, $\text{Var}(\mu) = \phi\mu$ (or $\text{Var}(\mu) = \phi\mu(1-\mu)$ for the binomial link).

The models are fit in the usual way. The parameter estimates are not affected by the value of ϕ. The covariance matrix, however, is multiplied by ϕ, and the scaled deviance and log likelihoods used in likelihood ratio tests are divided by ϕ.

The SCALE= option in the MODEL statement enables you to specify a value of ϕ for the Poisson (and also binomial) distributions. If you specify the SCALE=DEVIANCE option in the MODEL statement, the procedure uses the deviance divided by the degrees of freedom as an estimate of ϕ, and all statistics are adjusted appropriately. You can use Pearson's chi-square instead of the deviance by specifying the SCALE=PEARSON option.

```
run;
```
Executes the program

Many other options are available. See the SAS on-line help for further discussion and examples.

12.1
(a) We use SAS GENMOD to estimate the Poisson regression model with link
$$\ln\mu = \beta_0 + \beta_1 \ln(\text{MS}) + \beta_2 X2 + \ldots + \beta_5 X5 + \beta_6 Z2 + \ldots + \beta_8 Z4 + \beta_9 W2$$
Here X1 through X5 are the indicator variables for the type of ship (a class variable with five possibilities), Z1 through Z4 are the indicator variables for the year of construction (a class variable with four possibilities), and W1 and W2 are the indicator variables for the period of operation (a class variable with two possibilities). SAS GENMOD creates the associated indicator variables for the specified class variables automatically. The first outcome is declared as the reference.

The (type 3) test statistics at the end of the program output test the significance of the class variables. For example, the test statistic for "type" is obtained by comparing the log-likelihood of the full model (768.4585) with the log-likelihood of the restricted model that is missing that factor (the model with year, period, and ln(MS)). The log-likelihood of the restricted model is 762.1757. Hence the log-likelihood statistic is 2(768.4582 - 762.1757) = 12.57. Comparing this value to a chi-square with 4 degrees of freedom (since there are 4 restrictions), leads to the probability value P($\chi^2(4) \geq 12.57$) = 0.0136 . These are the values given at the end of the output. The tests for the other factors can be obtained similarly. They indicate that one can not simplify the model. All three factors are needed to explain the number of damage claims.

Ships of type 3 report the smallest number of damage incidents. Ships constructed in years 2 (1965-1969) and 3 (1970-1974) experience the highest number of reported damage incidents. The second period of operation (1975-79) is associated with a higher number of reported damage incidents.

<u>Fitting results for the full model:</u>

```
                        The GENMOD Procedure

                        Model Information

              Data Set              WORK.EXER12N1
              Distribution                Poisson
              Link Function                   Log
              Dependent Variable         nudamage
              Observations Used               34
                  Class Level Information

         Class     Value        Design Variables

         type        1        0    0    0    0
                     2        1    0    0    0
                     3        0    1    0    0
                     4        0    0    1    0
                     5        0    0    0    1


         year        1        0    0    0
                     2        1    0    0
                     3        0    1    0
                     4        0    0    1


         period      1        0
                     2        1


                     Parameter Information

     Parameter       Effect      type    year    period

     Prm1            Intercept
     Prm2            lnms
     Prm3            type         2
     Prm4            type         3
     Prm5            type         4
     Prm6            type         5
     Prm7            year                  2
     Prm8            year                  3
     Prm9            year                  4
     Prm10           period                         2
```

103

Criterion	DF	Value	Value/DF
Deviance	24	37.8043	1.5752
Scaled Deviance	24	37.8043	1.5752
Pearson Chi-Square	24	39.4494	1.6437
Scaled Pearson X2	24	39.4494	1.6437
Log Likelihood		768.4585	

Algorithm converged.

Estimated Correlation Matrix

	Prm1	Prm2	Prm3	Prm4	Prm5	Prm6	Prm7	Prm8	Prm9	Prm10
Prm1	1.0000	-0.9688	0.6048	-0.3172	-0.3046	-0.3304	-0.3405	-0.4538	-0.4298	-0.1729
Prm2	-0.9688	1.0000	-0.7587	0.2328	0.2200	0.2234	0.2291	0.3364	0.3495	0.1216
Prm3	0.6048	-0.7587	1.0000	0.0990	0.1226	0.1958	-0.1165	-0.0967	-0.1341	-0.0768
Prm4	-0.3172	0.2328	0.0990	1.0000	0.2798	0.3483	0.0899	0.1225	0.1660	0.0258
Prm5	-0.3046	0.2200	0.1226	0.2798	1.0000	0.3706	0.0788	0.1001	0.0024	0.0225
Prm6	-0.3304	0.2234	0.1958	0.3483	0.3706	1.0000	0.0466	0.0428	0.1200	0.0522
Prm7	-0.3405	0.2291	-0.1165	0.0899	0.0788	0.0466	1.0000	0.6612	0.5146	-0.0770
Prm8	-0.4538	0.3364	-0.0967	0.1225	0.1001	0.0428	0.6612	1.0000	0.5938	-0.1854
Prm9	-0.4298	0.3495	-0.1341	0.1660	0.0024	0.1200	0.5146	0.5938	1.0000	-0.2444
Prm10	-0.1729	0.1216	-0.0768	0.0258	0.0225	0.0522	-0.0770	-0.1854	-0.2444	1.0000

Analysis Of Parameter Estimates

Parameter		DF	Estimate	Standard Error	Wald 95% Confidence Limits		Chi-Square	Pr > ChiSq
Intercept		1	-5.5940	0.8724	-7.3038	-3.8841	41.12	<.0001
lnms		1	0.9027	0.1018	0.7032	1.1022	78.63	<.0001
type	2	1	-0.3499	0.2702	-0.8795	0.1797	1.68	0.1954
type	3	1	-0.7631	0.3382	-1.4259	-0.1003	5.09	0.0240
type	4	1	-0.1355	0.2971	-0.7178	0.4469	0.21	0.6484
type	5	1	0.2739	0.2418	-0.1999	0.7478	1.28	0.2572
year	2	1	0.6625	0.1536	0.3614	0.9637	18.60	<.0001
year	3	1	0.7597	0.1777	0.4115	1.1079	18.29	<.0001
year	4	1	0.3697	0.2458	-0.1121	0.8516	2.26	0.1326
period	2	1	0.3703	0.1181	0.1387	0.6018	9.82	0.0017
Scale		0	1.0000	0.0000	1.0000	1.0000		

LR Statistics For Type 3 Analysis

Source	DF	Chi-Square	Pr > ChiSq
lnms	1	101.28	<.0001
type	4	12.57	0.0136
year	3	27.20	<.0001
period	1	9.97	0.0016

Fitting results for the restricted model without type of ship:

The GENMOD Procedure

Model Information

Data Set	WORK.EXER12N1
Distribution	Poisson
Link Function	Log
Dependent Variable	nudamage
Observations Used	34

```
                     Class Level Information

              Class      Value     Design Variables

              year        1         0    0    0
                          2         1    0    0
                          3         0    1    0
                          4         0    0    1

              period      1         0
                          2         1

               Criteria For Assessing Goodness Of Fit

        Criterion                DF        Value      Value/DF

        Deviance                 28       50.3699      1.7989
        Scaled Deviance          28       50.3699      1.7989
        Pearson Chi-Square       28       46.7116      1.6683
        Scaled Pearson X2        28       46.7116      1.6683
        Log Likelihood                   762.1757

     Algorithm converged.

                   Analysis Of Parameter Estimates

                                  Standard   Wald 95% Confidence    Chi-
     Parameter      DF   Estimate   Error        Limits           Square   Pr > ChiSq

     Intercept      1    -5.2229   0.4826    -6.1688   -4.2771    117.12    <.0001
     lnms           1     0.8311   0.0460     0.7409    0.9213    326.13    <.0001
     year      2    1     0.6735   0.1503     0.3790    0.9681     20.08    <.0001
     year      3    1     0.7967   0.1702     0.4631    1.1303     21.91    <.0001
     year      4    1     0.3978   0.2337    -0.0603    0.8560      2.90    0.0887
     period    2    1     0.3546   0.1168     0.1256    0.5837      9.21    0.0024
     Scale          0     1.0000   0.0000     1.0000    1.0000

     NOTE: The scale parameter was held fixed.
```

(b) It seems reasonable to suppose that the number of damage incidents is directly proportional to MS, the months of service, and one can expect that the coefficient β_1 is one. The literature refers to the term $\ln(MS)$ as an "offset." Let us test for the offset, and test whether $\beta_1 = 1$. The estimate is $\hat{\beta}_1 = 0.9027$, and the 95 percent Wald confidence interval is given by $0.9027 \pm (1.96)(0.1018)$, 0.90 ± 0.20, or $0.70 \le \beta_1 \le 1.10$. The interval includes one, which makes the off-set interpretation plausible.

(c) We assume an "offset" for aggregate months of service (that is, we impose the restriction $\beta_1 = 1$) and estimate the model with link
$$\ln \mu = \beta_0 + \ln(MS) + \beta_2 X2 + \ldots + \beta_5 X5 + \beta_6 Z2 + \ldots + \beta_8 Z4 + \beta_9 W2$$
The results of the estimation are similar to the ones of the full model in (a).

Fitting results for the model with an offset:

```
                        The GENMOD Procedure

                        Model Information

           Data Set               WORK.EXER12N1
           Distribution                 Poisson
           Link Function                    Log
           Dependent Variable          nudamage
           Offset Variable                 lnms
           Observations Used                 34

                    Class Level Information

          Class     Value        Design Variables

          type        1        0    0    0    0
                      2        1    0    0    0
                      3        0    1    0    0
                      4        0    0    1    0
                      5        0    0    0    1

          year        1        0    0    0
                      2        1    0    0
                      3        0    1    0
                      4        0    0    1

          period      1        0
                      2        1

                    Parameter Information

          Parameter    Effect      type    year    period

          Prm1         Intercept
          Prm2         type         2
          Prm3         type         3
          Prm4         type         4
          Prm5         type         5
          Prm6         year                  2
          Prm7         year                  3
          Prm8         year                  4
          Prm9         period                         2

              Criteria For Assessing Goodness Of Fit

          Criterion           DF        Value      Value/DF

          Deviance            25      38.6951        1.5478
          Scaled Deviance     25      38.6951        1.5478
          Pearson Chi-Square  25      42.2753        1.6910
          Scaled Pearson X2   25      42.2753        1.6910
          Log Likelihood              768.0131

      Algorithm converged.

                    Estimated Correlation Matrix

        Prm1     Prm2     Prm3     Prm4     Prm5     Prm6     Prm7     Prm8     Prm9

Prm1   1.0000  -0.8114  -0.3784  -0.3706  -0.4699  -0.4843  -0.5501  -0.4015  -0.2161
Prm2  -0.8114   1.0000   0.4332   0.4468   0.5707   0.0856   0.2714   0.2285   0.0254
Prm3  -0.3784   0.4332   1.0000   0.2375   0.3136   0.0358   0.0455   0.0971  -0.0031
Prm4  -0.3706   0.4468   0.2375   1.0000   0.3338   0.0277   0.0286  -0.0966  -0.0047
Prm5  -0.4699   0.5707   0.3136   0.3338   1.0000  -0.0041  -0.0371   0.0528   0.0269
```

```
Prm6    -0.4843   0.0856   0.0358   0.0277   -0.0041    1.0000   0.6335   0.4755   -0.1201
Prm7    -0.5501   0.2714   0.0455   0.0286   -0.0371    0.6335   1.0000   0.5482   -0.2636
Prm8    -0.4015   0.2285   0.0971  -0.0966    0.0528    0.4755   0.5482   1.0000   -0.3154
Prm9    -0.2161   0.0254  -0.0031  -0.0047    0.0269   -0.1201  -0.2636  -0.3154    1.0000
```

Analysis Of Parameter Estimates

Parameter	DF	Estimate	Standard Error	Wald 95% Confidence Limits		Chi-Square	Pr > ChiSq
Intercept		1	-6.4059	0.2174	-6.8321	-5.9797	867.89	<.0001
type	2	1	-0.5433	0.1776	-0.8914	-0.1953	9.36	0.0022
type	3	1	-0.6874	0.3290	-1.3323	-0.0425	4.36	0.0367
type	4	1	-0.0760	0.2906	-0.6455	0.4936	0.07	0.7938
type	5	1	0.3256	0.2359	-0.1367	0.7879	1.91	0.1675
year	2	1	0.6971	0.1496	0.4038	0.9904	21.70	<.0001
year	3	1	0.8184	0.1698	0.4857	1.1512	23.24	<.0001
year	4	1	0.4534	0.2332	-0.0036	0.9104	3.78	0.0518
period	2	1	0.3845	0.1183	0.1527	0.6163	10.57	0.0012
Scale		0	1.0000	0.0000	1.0000	1.0000		

NOTE: The scale parameter was held fixed.

LR Statistics For Type 3 Analysis

Source	DF	Chi-Square	Pr > ChiSq
type	4	23.67	<.0001
year	3	31.41	<.0001
period	1	10.66	0.0011

(d) Let us look at the deviance goodness-of-fit statistics. Comparing the deviance $D = 37.8043$ to a chi-square with 24 degrees of freedom, leads to the probability value $P(\chi^2(24) \geq 37.80) = 1 - 0.9637 = 0.0363$. The deviance exceeds the 95th percentile and the probability value is slightly smaller than 0.05. This is a sign of overdispersion. We adjust the analysis for overdispersion by allowing the variance function of the Poisson distribution to have a multplicative overdispersion factor, $\text{Var}(\mu) = \phi\mu$. The model is fit in the usual way, and the parameter estimates are not affected by the value of ϕ. The covariance matrix, however, is multiplied by ϕ, and the scaled deviance and log likelihoods used in likelihood ratio tests are divided by ϕ. The SCALE=DEVIANCE option in the MODEL statement enables us to specify a value of ϕ for the Poisson distribution. The procedure uses the deviance divided by the degrees of freedom as an estimate of ϕ, and all statistics are adjusted appropriately.

The results are basically unchanged. The test statistics indicate that all three factors are statistically significant. Ships of types 2 and 3 experience the smallest numbers of reported damage incidents. Ships constructed in years 2 (1965-1969) and 3 (1970-1974) experience the largest numbers of reported damage incidents. The second period of operation (1975-79) is associated with a higher number of reported damage incidents.

Fitting results for the model with scale adjustment:

```
                        The GENMOD Procedure

                        Model Information

              Data Set              WORK.EXER12N1
              Distribution                Poisson
              Link Function                   Log
              Dependent Variable         nudamage
              Offset Variable                lnms
              Observations Used                34
```

Class Level Information

Class	Value	Design Variables			
type	1	0	0	0	0
	2	1	0	0	0
	3	0	1	0	0
	4	0	0	1	0
	5	0	0	0	1
year	1	0	0	0	
	2	1	0	0	
	3	0	1	0	
	4	0	0	1	
period	1	0			
	2	1			

Parameter Information

Parameter	Effect	type	year	period
Prm1	Intercept			
Prm2	type	2		
Prm3	type	3		
Prm4	type	4		
Prm5	type	5		
Prm6	year		2	
Prm7	year		3	
Prm8	year		4	
Prm9	period			2

Criteria For Assessing Goodness Of Fit

Criterion	DF	Value	Value/DF
Deviance	25	38.6951	1.5478
Scaled Deviance	25	25.0000	1.0000
Pearson Chi-Square	25	42.2753	1.6910
Scaled Pearson X2	25	27.3131	1.0925
Log Likelihood		496.1960	

Algorithm converged.

Estimated Correlation Matrix

	Prm1	Prm2	Prm3	Prm4	Prm5	Prm6	Prm7	Prm8	Prm9
Prm1	1.0000	-0.8114	-0.3784	-0.3706	-0.4699	-0.4843	-0.5501	-0.4015	-0.2161
Prm2	-0.8114	1.0000	0.4332	0.4468	0.5707	0.0856	0.2714	0.2285	0.0254
Prm3	-0.3784	0.4332	1.0000	0.2375	0.3136	0.0358	0.0455	0.0971	-0.0031
Prm4	-0.3706	0.4468	0.2375	1.0000	0.3338	0.0277	0.0286	-0.0966	-0.0047
Prm5	-0.4699	0.5707	0.3136	0.3338	1.0000	-0.0041	-0.0371	0.0528	0.0269

```
Prm6    -0.4843   0.0856   0.0358   0.0277  -0.0041   1.0000   0.6335   0.4755  -0.1201
Prm7    -0.5501   0.2714   0.0455   0.0286  -0.0371   0.6335   1.0000   0.5482  -0.2636
Prm8    -0.4015   0.2285   0.0971  -0.0966   0.0528   0.4755   0.5482   1.0000  -0.3154
Prm9    -0.2161   0.0254  -0.0031  -0.0047   0.0269  -0.1201  -0.2636  -0.3154   1.0000
```

```
                          Analysis Of Parameter Estimates

                                 Standard   Wald 95% Confidence    Chi-
Parameter        DF   Estimate     Error         Limits          Square    Pr > ChiSq

Intercept         1    -6.4059    0.2705    -6.9361    -5.8757    560.72      <.0001
type      2       1    -0.5433    0.2209    -0.9764    -0.1103      6.05      0.0139
type      3       1    -0.6874    0.4094    -1.4898     0.1149      2.82      0.0931
type      4       1    -0.0760    0.3615    -0.7845     0.6326      0.04      0.8336
type      5       1     0.3256    0.2935    -0.2496     0.9007      1.23      0.2672
year      2       1     0.6971    0.1862     0.3323     1.0620     14.02      0.0002
year      3       1     0.8184    0.2112     0.4044     1.2324     15.01      0.0001
year      4       1     0.4534    0.2901    -0.1151     1.0220      2.44      0.1180
period    2       1     0.3845    0.1471     0.0961     0.6729      6.83      0.0090
Scale             0     1.2441    0.0000     1.2441     1.2441
```

NOTE: The scale parameter was estimated by the square root of DEVIANCE/DOF.

```
                    LR Statistics For Type 3 Analysis
                                                   Chi-
        Source      Num DF   Den DF   F Value   Pr > F   Square   Pr > ChiSq

        type          4       25      3.82     0.0147   15.29      0.0041
        year          3       25      6.76     0.0017   20.29      0.0001
        period        1       25      6.89     0.0146    6.89      0.0087
```

(e) A model with every possible two-factor interaction contains
1 (const) + 4 + 3 + 1 (main effects) + 4*3 + 4*1 + 3*1 (2-factor interactions) = 28
parameters. This is a highly non-parsimonious model, considering that there are only
34 observations. The number of parameters in the fully saturated model (with the 3-
factor interaction added) exceeds the number of observations.

Here we enter each two-factor interaction one at-a-time. The type 3 test results for the
models with the type by period interaction (4 additional parameters) and the year by
period interaction (3 additional parameters) are given below. The model with the type
by year interaction (12 additional parameters) experienced convergence problems,
probably due to the large number of additional parameters and the sparseness of the
data. The results indicate that interaction components are not needed. Note that type 3
LR test statistics are partial tests, always testing whether the factor in question is
significant when added last to the model. The period effect is insignificant when
adding it to the model with type, year, and the type by period interaction. However, it
becomes significant when the type by period interaction is omitted.

Fitting results for the model with interaction:

```
                    LR Statistics For Type 3 Analysis

                                    Chi-
               Source         DF    Square    Pr > ChiSq
```

109

type	4	12.13	0.0164
year	3	30.70	<.0001
period	1	1.57	0.2105
type*period	4	4.94	0.2936

LR Statistics For Type 3 Analysis

Source	DF	Chi-Square	Pr > ChiSq
type	4	23.71	<.0001
year	3	25.26	<.0001
period	1	7.29	0.0069
year*period	3	4.00	0.2613

(f) See parts (a) – (e)

12.4

(a) Cancer incidence should be directly proportional to the size of the population. Hence it is reasonable to consider ln(POP) as an offset. Age is a categorical variable. We use indicator variables for the eight age groups (X1 through X8) and consider the Poisson regression with link

$$\ln \mu = \beta_0 + \ln(POP) + \beta_2 X2 + \dots + \beta_8 X8 + \beta_9 \text{Town}$$

The results of the model fit are shown below. Both age and town are significant; you can see this from the (partial; type 3) likelihood-ratio test statistics and their probability values at the end of the output. The estimate of the town effect is $\hat{\beta}_9 = 0.85$, with standard error 0.06. There is a significant location effect; women in Texas have a $100[\exp(0.85) - 1] = 134$ percent higher incidence of skin cancer.

The deviance and the Pearson Chi-Square statistics are approximately one and indicate no problem with over/under-dispersion.

Fitting results for the full model with an offset:

The GENMOD Procedure

Model Information

Data Set	WORK.EXER12N4
Distribution	Poisson
Link Function	Log
Dependent Variable	nucases
Offset Variable	lnpop
Observations Used	15

Class Level Information

Class	Value	Design Variables

110

```
age    1        0    0    0    0    0    0    0
       2        1    0    0    0    0    0    0
       3        0    1    0    0    0    0    0
       4        0    0    1    0    0    0    0
       5        0    0    0    1    0    0    0
       6        0    0    0    0    1    0    0
       7        0    0    0    0    0    1    0
       8        0    0    0    0    0    0    1
```

Parameter Information

Parameter	Effect	age
Prm1	Intercept	
Prm2	town	
Prm3	age	2
Prm4	age	3
Prm5	age	4
Prm6	age	5
Prm7	age	6
Prm8	age	7
Prm9	age	8

Criteria For Assessing Goodness Of Fit

Criterion	DF	Value	Value/DF
Deviance	6	5.2089	0.8682
Scaled Deviance	6	5.2089	0.8682
Pearson Chi-Square	6	5.1482	0.8580
Scaled Pearson X2	6	5.1482	0.8580
Log Likelihood		6204.3156	

Estimated Correlation Matrix

	Prm1	Prm2	Prm3	Prm4	Prm5	Prm6	Prm7	Prm8	Prm9
Prm1	1.0000	-0.0944	-0.9521	-0.9788	-0.9868	-0.9885	-0.9900	-0.9819	-0.9730
Prm2	-0.0944	1.0000	-0.0031	-0.0047	-0.0037	-0.0024	0.0007	0.0927	0.0039
Prm3	-0.9521	-0.0031	1.0000	0.9410	0.9486	0.9501	0.9513	0.9349	0.9347
Prm4	-0.9788	-0.0047	0.9410	1.0000	0.9753	0.9769	0.9781	0.9610	0.9610
Prm5	-0.9868	-0.0037	0.9486	0.9753	1.0000	0.9847	0.9860	0.9689	0.9687
Prm6	-0.9885	-0.0024	0.9501	0.9769	0.9847	1.0000	0.9875	0.9706	0.9703
Prm7	-0.9900	0.0007	0.9513	0.9781	0.9860	0.9875	1.0000	0.9721	0.9715
Prm8	-0.9819	0.0927	0.9349	0.9610	0.9689	0.9706	0.9721	1.0000	0.9554
Prm9	-0.9730	0.0039	0.9347	0.9610	0.9687	0.9703	0.9715	0.9554	1.0000

Analysis Of Parameter Estimates

Parameter		DF	Estimate	Standard Error	Wald 95% Confidence Limits		Chi-Square	Pr > ChiSq
Intercept		1	-11.6921	0.4492	-12.5725	-10.8116	677.43	<.0001
town		1	0.8527	0.0596	0.7358	0.9696	204.54	<.0001
age	2	1	2.6290	0.4675	1.7128	3.5452	31.63	<.0001
age	3	1	3.8456	0.4547	2.9545	4.7367	71.54	<.0001
age	4	1	4.5938	0.4510	3.7098	5.4778	103.74	<.0001
age	5	1	5.0864	0.4503	4.2038	5.9690	127.59	<.0001
age	6	1	5.6457	0.4497	4.7642	6.5272	157.58	<.0001
age	7	1	6.2032	0.4575	5.3065	7.0999	183.83	<.0001
age	8	1	6.1757	0.4577	5.2785	7.0728	182.02	<.0001
Scale		0	1.0000	0.0000	1.0000	1.0000		

```
                 LR Statistics For Type 3 Analysis

                                   Chi-
            Source          DF    Square   Pr > ChiSq

            town             1    226.52     <.0001
            age              7   2199.01     <.0001
```

(b) The estimation results for the more general model
$$\ln \mu = \beta_0 + \beta_1 \ln(POP) + \beta_2 X2 + ... + \beta_8 X8 + \beta_9 \text{Town}$$
are given below. It seems reasonable to suppose that the number of cancers is directly proportional to the population, and that one can expect that the coefficient β_1 is one. Let us test whether $\beta_1 = 1$. The estimate is $\hat{\beta}_1 = 1.96$, and the 95 percent Wald confidence interval is given by $1.96 \pm (1.96)(0.63)$, 1.96 ± 1.23, or $0.73 \le \beta_1 \le 3.18$. The interval is quite wide (because there are few observations). The interval includes one, which makes the off-set interpretation plausible.

Fitting results for the full model without an offset:

```
                       The GENMOD Procedure

                         Model Information

              Data Set              WORK.EXER12N4
              Distribution                Poisson
              Link Function                   Log
              Dependent Variable          nucases
              Observations Used                15

                    Class Level Information

    Class   Value               Design Variables

    age       1      0    0   0    0    0    0   0
              2      1    0   0    0    0    0   0
              3      0    1   0    0    0    0   0
              4      0    0   1    0    0    0   0
              5      0    0   0    1    0    0   0
              6      0    0   0    0    1    0   0
              7      0    0   0    0    0    1   0
              8      0    0   0    0    0    0   1

                    Parameter Information

             Parameter     Effect      age

             Prm1          Intercept
             Prm2          lnpop
             Prm3          town
             Prm4          age          2
             Prm5          age          3
             Prm6          age          4
             Prm7          age          5
             Prm8          age          6
```

112

```
                     Prm9          age        7
                     Prm10         age        8

                Criteria For Assessing Goodness Of Fit

        Criterion              DF       Value      Value/DF

        Deviance                5      2.8539       0.5708
        Scaled Deviance         5      2.8539       0.5708
        Pearson Chi-Square      5      2.8439       0.5688
        Scaled Pearson X2       5      2.8439       0.5688
        Log Likelihood                 6205.4931
```

                    The GENMOD Procedure

Algorithm converged.

                Estimated Correlation Matrix

```
        Prm1     Prm2     Prm3     Prm4     Prm5     Prm6     Prm7     Prm8     Prm9     Prm10

Prm1    1.0000  -0.9982   0.7154  -0.3692  -0.5729  -0.6353  -0.7844  -0.8810  -0.9360  -0.9851
Prm2   -0.9982   1.0000  -0.7206   0.3160   0.5241   0.5888   0.7465   0.8516   0.9138   0.9736
Prm3    0.7154  -0.7206   1.0000  -0.2317  -0.3831  -0.4284  -0.5401  -0.6131  -0.6327  -0.7003
Prm4   -0.3692   0.3160  -0.2317   1.0000   0.9260   0.9135   0.8357   0.7422   0.6489   0.5100
Prm5   -0.5729   0.5241  -0.3831   0.9260   1.0000   0.9800   0.9448   0.8830   0.8112   0.6971
Prm6   -0.6353   0.5888  -0.4284   0.9135   0.9800   1.0000   0.9691   0.9192   0.8560   0.7520
Prm7   -0.7844   0.7465  -0.5401   0.8357   0.9448   0.9691   1.0000   0.9802   0.9444   0.8742
Prm8   -0.8810   0.8516  -0.6131   0.7422   0.8830   0.9192   0.9802   1.0000   0.9852   0.9453
Prm9   -0.9360   0.9138  -0.6327   0.6489   0.8112   0.8560   0.9444   0.9852   1.0000   0.9783
Prm10  -0.9851   0.9736  -0.7003   0.5100   0.6971   0.7520   0.8742   0.9453   0.9783   1.0000
```

                   Analysis Of Parameter Estimates

```
                            Standard   Wald 95% Confidence    Chi-
Parameter        DF  Estimate  Error        Limits          Square   Pr > ChiSq

Intercept        1  -23.2489  7.5392  -38.0256   -8.4723      9.51      0.0020
lnpop            1    1.9613  0.6259    0.7345    3.1880      9.82      0.0017
town             1    0.7556  0.0862    0.5866    0.9245     76.81     <.0001
age        2     1    2.8684  0.4927    1.9027    3.8341     33.89     <.0001
age        3     1    4.2766  0.5339    3.2303    5.3230     64.17     <.0001
age        4     1    5.0990  0.5580    4.0053    6.1927     83.49     <.0001
age        5     1    5.8623  0.6768    4.5358    7.1888     75.02     <.0001
age        6     1    6.7681  0.8579    5.0866    8.4496     62.23     <.0001
age        7     1    7.7827  1.1265    5.5748    9.9906     47.73     <.0001
age        8     1    9.1783  2.0057    5.2473   13.1094     20.94     <.0001
Scale            0    1.0000  0.0000    1.0000    1.0000
```

NOTE: The scale parameter was held fixed.

                 LR Statistics For Type 3 Analysis

```
                            Chi-
        Source        DF   Square    Pr > ChiSq

        lnpop          1     9.77      0.0018
        town           1    81.60     <.0001
        age            7   988.50     <.0001
```

Additional model: We estimate a model that includes an interaction between town and age. We want to check whether the town effect depends on the age group. The results are given below. The likelihood-ratio test for the town by age interaction is insignificant. Note that such a test is possible in the saturated Poisson regression model, as the variance is the same as the mean; the scale parameter is kept fixed.

Fitting results for the model with interaction:

```
                    The GENMOD Procedure

                    Model Information

            Data Set             WORK.EXER12N4
            Distribution              Poisson
            Link Function                 Log
            Dependent Variable        nucases
            Offset Variable             lnpop
            Observations Used              15

                Class Level Information

Class     Value                Design Variables

age       1      0    0    0    0    0    0    0
          2      1    0    0    0    0    0    0
          3      0    1    0    0    0    0    0
          4      0    0    1    0    0    0    0
          5      0    0    0    1    0    0    0
          6      0    0    0    0    1    0    0
          7      0    0    0    0    0    1    0
          8      0    0    0    0    0    0    1
town      0      0    0
          1      1    1

                 Parameter Information

            Parameter    Effect        age    town

            Prm1         Intercept
            Prm2         town                   1
            Prm3         age            2
            Prm4         age            3
            Prm5         age            4
            Prm6         age            5
            Prm7         age            6
            Prm8         age            7
            Prm9         age            8
            Prm10        age*town       2       1
            Prm11        age*town       3       1
            Prm12        age*town       4       1
            Prm13        age*town       5       1
            Prm14        age*town       6       1
            Prm15        age*town       7       1
            Prm16        age*town       8       1

            Criteria For Assessing Goodness Of Fit

    Criterion              DF        Value       Value/DF

    Deviance                0       0.0000          .
    Scaled Deviance         0       0.0000          .
    Pearson Chi-Square      0       0.0000          .
    Scaled Pearson X2       0       0.0000          .
    Log Likelihood               6206.9201

Algorithm converged.
```

114

Parameter			DF	Estimate	Standard Error	Wald 95% Confidence Limits		Chi-Square	Pr > ChiSq
Intercept			1	-12.0592	1.0000	-14.0191	-10.0992	145.42	<.0001
town	1		1	1.3373	1.1180	-0.8540	3.5286	1.43	0.2316
age	2		1	3.1113	1.0308	1.0910	5.1316	9.11	0.0025
age	3		1	3.9860	1.0165	1.9937	5.9784	15.38	<.0001
age	4		1	4.8917	1.0070	2.9180	6.8655	23.60	<.0001
age	5		1	5.4975	1.0049	3.5280	7.4671	29.93	<.0001
age	6		1	6.0167	1.0038	4.0492	7.9842	35.92	<.0001
age	7		1	6.5703	1.0038	4.6029	8.5376	42.85	<.0001
age	8		1	6.7207	1.0124	4.7364	8.7050	44.07	<.0001
age*town	2	1	1	-0.6446	1.1571	-2.9124	1.6232	0.31	0.5774
age*town	3	1	1	-0.1917	1.1365	-2.4193	2.0359	0.03	0.8661
age*town	4	1	1	-0.3922	1.1263	-2.5998	1.8154	0.12	0.7277
age*town	5	1	1	-0.5455	1.1241	-2.7487	1.6578	0.24	0.6275
age*town	6	1	1	-0.4901	1.1229	-2.6910	1.7107	0.19	0.6625
age*town	7	1	0	0.0000	0.0000	0.0000	0.0000	.	.
age*town	8	1	1	-0.7581	1.1360	-2.9845	1.4683	0.45	0.5045
Scale			0	1.0000	0.0000	1.0000	1.0000		

NOTE: The scale parameter was held fixed.

LR Statistics For Type 3 Analysis

Source	DF	Chi-Square	Pr > ChiSq
town	1	1.78	0.1817
age	7	845.79	<.0001
age*town	7	5.21	0.6342

Another model: Finally, we introduce age as a continuous variable, and not as a factor as was done in the previous models. The output is shown below. Both age and town are significant. A graph of the number of cancer deaths against age (with the two towns indicated by different plotting symbols) and the Poisson model fit is given in the following graph. Every ten years the cancer rate (deaths per population) increases by a factor of $\exp(0.6133) = 1.85$; that is, by 85 percent.

Fitting results for the model with age as continuous variable:

The GENMOD Procedure

Model Information

Data Set	WORK.EXER12N4
Distribution	Poisson
Link Function	Log
Dependent Variable	nucases
Offset Variable	lnpop
Observations Used	15

Criteria For Assessing Goodness Of Fit

Criterion	DF	Value	Value/DF
Deviance	12	184.8091	15.4008
Scaled Deviance	12	184.8091	15.4008

```
            Pearson Chi-Square        12        141.4307        11.7859
            Scaled Pearson X2         12        141.4307        11.7859
            Log Likelihood                      6114.5155
```

Algorithm converged.

```
                        Analysis Of Parameter Estimates

                            Standard    Wald 95% Confidence      Chi-
     Parameter   DF   Estimate   Error        Limits           Square   Pr > ChiSq

     Intercept    1   -9.8191   0.0902   -9.9959    -9.6423    11846.5     <.0001
     town         1    0.8584   0.0545    0.7515     0.9652     247.95     <.0001
     age          1    0.6133   0.0142    0.5855     0.6411    1871.42     <.0001
     Scale        0    1.0000   0.0000    1.0000     1.0000
```

NOTE: The scale parameter was held fixed.

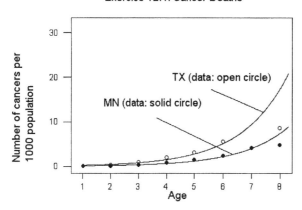

Exercise 12.4: Cancer Deaths

12.6 The output from estimating the Poisson regression with link
$$\ln \mu = \beta_0 + \beta_1 \text{DIST} + \beta_2 \text{INC} + \beta_3 \text{SIZE2} + \beta_4 \text{SIZE3} + \beta_5 \text{SIZE4} + \beta_6 \text{SIZE5}$$
is shown below. Here we treat SIZE as a class variable, specifying 4 indicators for the factor with five outcomes (1 through 5 people; size 1 is the baseline). Income does not affect the number of visits to the lake (probability value = 0.27) and is omitted in the next run. The deviance and the Pearson chi-square statistics are roughly the size of the critical 95th percentile (280.36).

Fitting results for the full model:

```
                        The GENMOD Procedure

                        Model Information

            Data Set            WORK.EXER12N6
            Distribution              Poisson
            Link Function                 Log
```

116

```
                    Dependent Variable         nuvisits
                    Observations Used              250

                    Class Level Information

          Class    Value       Design Variables

          size     1        0      0      0      0
                   2        1      0      0      0
                   3        0      1      0      0
                   4        0      0      1      0
                   5        0      0      0      1


                    Parameter Information

          Parameter       Effect       size

          Prm1            Intercept
          Prm2            dist
          Prm3            inc
          Prm4            size          2
          Prm5            size          3
          Prm6            size          4
          Prm7            size          5

              Criteria For Assessing Goodness Of Fit

     Criterion              DF        Value       Value/DF

     Deviance               243      313.6999      1.2909
     Scaled Deviance        243      313.6999      1.2909
     Pearson Chi-Square     243      286.2022      1.1778
     Scaled Pearson X2      243      286.2022      1.1778
     Log Likelihood                   11.3651

  Algorithm converged.

                    Estimated Correlation Matrix

            Prm1       Prm2       Prm3       Prm4       Prm5       Prm6       Prm7

  Prm1     1.0000    -0.4081    -0.6605    -0.4409    -0.5760    -0.5506    -0.5740
  Prm2    -0.4081     1.0000    -0.0275     0.0247     0.1431     0.0573     0.0438
  Prm3    -0.6605    -0.0275     1.0000    -0.0391     0.0536     0.0374     0.0749
  Prm4    -0.4409     0.0247    -0.0391     1.0000     0.5739     0.5990     0.6019
  Prm5    -0.5760     0.1431     0.0536     0.5739     1.0000     0.6386     0.6437
  Prm6    -0.5506     0.0573     0.0374     0.5990     0.6386     1.0000     0.6678
  Prm7    -0.5740     0.0438     0.0749     0.6019     0.6437     0.6678     1.0000


  Analysis Of Parameter Estimates

                            Standard    Wald 95% Confidence      Chi-
  Parameter      DF  Estimate   Error        Limits            Square   Pr > ChiSq

  Intercept      1    1.6578   0.1907    1.2840    2.0315       75.57     <.0001
  dist           1   -0.0215   0.0016   -0.0245   -0.0184      190.19     <.0001
  inc            1    0.0203   0.0184   -0.0158    0.0563        1.22      0.2700
  size      2    1   -0.0249   0.1595   -0.3375    0.2877        0.02      0.8758
  size      3    1    0.1032   0.1521   -0.1949    0.4014        0.46      0.4973
  size      4    1    0.3344   0.1454    0.0495    0.6194        5.29      0.0214
  size      5    1    0.4731   0.1442    0.1904    0.7558       10.76      0.0010
  Scale          0    1.0000   0.0000    1.0000    1.0000

  NOTE: The scale parameter was held fixed.
```

117

```
                    LR Statistics For Type 3 Analysis

                                    Chi-
                Source          DF  Square   Pr > ChiSq

                dist             1  213.97     <.0001
                inc              1    1.22     0.2699
                size             4   21.19     0.0003
```

The output of the simplified Poisson regression with link
$$\ln \mu = \beta_0 + \beta_1 \text{DIST} + \beta_3 \text{SIZE2} + \beta_4 \text{SIZE3} + \beta_5 \text{SIZE4} + \beta_6 \text{SIZE5}$$
is shown below.

Fitting results for the restricted model without income:

```
                        The GENMOD Procedure

                        Model Information

              Data Set              WORK.EXER12N6
              Distribution                Poisson
              Link Function                   Log
              Dependent Variable          nuvisits
              Observations Used               250

                    Class Level Information

        Class    Value        Design Variables

        size     1         0    0    0    0
                 2         1    0    0    0
                 3         0    1    0    0
                 4         0    0    1    0
                 5         0    0    0    1

              Criteria For Assessing Goodness Of Fit

        Criterion            DF        Value      Value/DF

        Deviance            244     314.9173        1.2906
        Scaled Deviance     244     314.9173        1.2906
        Pearson Chi-Square  244     284.7341        1.1669
        Scaled Pearson X2   244     284.7341        1.1669
        Log Likelihood              10.7564

    Algorithm converged.

                    Analysis Of Parameter Estimates

                              Standard  Wald 95% Confidence    Chi-
    Parameter       DF  Estimate  Error      Limits          Square   Pr > ChiSq

    Intercept        1   1.7957   0.1431   1.5152   2.0762   157.44     <.0001
    dist             1  -0.0214   0.0016  -0.0245  -0.0184   189.88     <.0001
    size       2     1  -0.0184   0.1594  -0.3308   0.2939     0.01     0.9079
    size       3     1   0.0941   0.1519  -0.2035   0.3917     0.38     0.5355
    size       4     1   0.3283   0.1453   0.0436   0.6130     5.11     0.0238
    size       5     1   0.4610   0.1439   0.1790   0.7429    10.27     0.0014
    Scale            0   1.0000   0.0000   1.0000   1.0000

NOTE: The scale parameter was held fixed.
```

118

Additional model: Treating SIZE as a continuous variable and not as a factor leads to the Poisson link

$$\ln \mu = \beta_0 + \beta_1 \mathrm{DIST} + \beta_2 \mathrm{INC} + \beta_3 \mathrm{SIZE}.$$

The estimation results show that income can be omitted (output not shown). Omitting income leads to the results shown below. Both distance and family size are statistically significant. A change in distance by 10 miles reduces the mean number of visits by a factor of $\exp(-0.0212(10)) = 0.81$, or 19 percent. A change in the family size by one unit increases the mean number of visits by a factor $\exp(0.1358) = 1.145$, or 14.5 percent.

We can test whether a class factor for size is needed or whether it is sufficient to treat size as a continuous variable. The log-likelihood of the model that considers size as a factor (the full model) is 10.7564; the log-likelihood of the model that considers size as a continuous variable (the restricted model) is 9.8849. We compare the log-likelihood ratio statistic, $2(10.7564-9.8849) = 1.74$, to a chi-square with 3 degrees of freedom (the five intercepts in the unrestricted model, one for each of the five size groups, are tested against the linear formulation which includes two parameters, the intercept and the slope). The test statistic is small (probability value $P(\chi^2(3) \geq 1.74) = 1 - 0.37 = 0.63$ is large), indicating that it is sufficient to consider a linear component of size. A scatter plot of the number of visits against distance and fitted values from the Poisson regression against distance is also shown.

Fitting results for the model with size as continuous variable:

```
                      The GENMOD Procedure

                       Model Information

            Data Set              WORK.EXER12N6
            Distribution                Poisson
            Link Function                   Log
            Dependent Variable          nuvisits
            Observations Used                250


    Criteria For Assessing Goodness Of Fit

         Criterion            DF        Value      Value/DF

         Deviance            247     316.6602        1.2820
         Scaled Deviance     247     316.6602        1.2820
         Pearson Chi-Square  247     287.1920        1.1627
         Scaled Pearson X2   247     287.1920        1.1627
         Log Likelihood                9.8849

    Algorithm converged.
```

```
                       Analysis Of Parameter Estimates

                               Standard    Wald 95% Confidence    Chi-
         Parameter    DF   Estimate   Error       Limits        Square   Pr > ChiSq

         Intercept    1    1.5453    0.1367    1.2774   1.8133   127.75    <.0001
         dist         1   -0.0212    0.0015   -0.0242  -0.0182   191.10    <.0001
         size         1    0.1358    0.0317    0.0736   0.1980    18.32    <.0001
         Scale        0    1.0000    0.0000    1.0000   1.0000

         NOTE: The scale parameter was held fixed.
```

Exercise 12.6

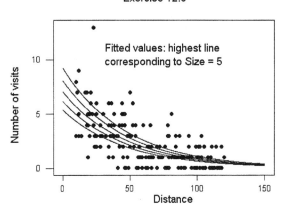

CPSIA information can be obtained
at www.ICGtesting.com
Printed in the USA
FFHW010148120619
52961463-58557FF